THE SCIENCE OF SUBTLE ENERGY

"Yury Kronn's *The Science of Subtle Energy* is a formidable work that should be read by both the general public and the scientific community. The general reader will be spellbound with tales of scientific discovery through the ages, and the scientist will once again be reminded that many, if not most, serious scientific breakthroughs have been met with derision by members of the academy, who protect their established 'truths' from threats of novel experiment and novel theory. *The Science of Subtle Energy* is a threatening book. Kronn has meticulously outlined both novel experiments and novel theory to grapple with perhaps the greatest scientific puzzle ever—what is the mysterious dark matter and dark energy that makes up the vast majority of the universe and about which we know essentially nothing? Kronn makes a compelling case that the study of 'subtle energies' provides the key. This isn't just conjecture. Herein he presents innumerable experimental data and a novel technology that deserve careful scrutiny and replication by others. For the intellectually curious, what might be the implications, both theoretical and practical, if even some of his work bears fruit? Wow."

William F. Bengston, Ph.D., president of the Society for Scientific Exploration

"*The Science of Subtle Energy* is both exciting and frustrating. The excitement comes from following the intellectual trail Yury Kronn has mapped from the early years of recognizing the existence of subtle energy and its potential positives and negatives to his own work demonstrating the practical applications of Vital Force Technology to his predictions of additional benefits to be obtained with further in vitro and in vivo research. The frustration stems from wanting to move faster to exploit the broad spectrum of potential indications of subtle energy, while simultaneously limiting deleterious effects that may well be manifested in our world of electronic contamination. *The Science of Subtle Energy* is a page-turner, an eye-opener, and an inducer of humbleness as we learn how little we know and how difficult it is to alter paradigms in the face of new information."

John McMichael, Ph.D., immunologist, virologist, and founder of Milkhaus Laboratory, Inc.

THE SCIENCE OF SUBTLE ENERGY

The Healing Power of Dark Matter

Yury Kronn, Ph.D., with Jurriaan Kamp

Park Street Press
Rochester, Vermont

Park Street Press
One Park Street
Rochester, Vermont 05767
www.ParkStPress.com

Text stock is SFI certified

Park Street Press is a division of Inner Traditions International

Cataloging-in-Publication Data for this title is available from the Library of Congress

ISBN 978-1-64411-452-0 (print)
ISBN 978-1-64411-453-7 (ebook)

Printed and bound in the United States by Lake Book Manufacturing, Inc.
The text stock is SFI certified. The Sustainable Forestry Initiative® program promotes sustainable forest management.

10 9 8 7 6 5 4 3 2 1

Text design and layout by Priscilla Harris Baker
This book was typeset in Garamond Premier Pro with Blacker, Futura, and Titular used as display typefaces

To send correspondence to the author of this book, mail a first-class letter to the author c/o Inner Traditions • Bear & Company, One Park Street, Rochester, VT 05767, and we will forward the communication.

To Constance Kronn (1948–2017)

"There are numberless energy rays in the Universe. . . . All rays . . . come from the subtle nature of the Universe. . . . The interwoven energy net influences the lives of individual human beings, whole societies and entire races. . . . A virtuous individual who responds to the high, pure, harmonious subtle energy rays and integrates them with the positive elements of his own inner being may strengthen his life, enhance his health and power, and lengthen his years."

—LAO-TZU, *HUA HU CHING,* 500 BCE

Contents

FOREWORD

Predictable Observations Are No Accidents

Bruce Lipton, Ph.D.

MUCH OF SCIENCE, including medicine and biology, is based on Newtonian physics established in the second half of the seventeenth century. Isaac Newton sought to understand the mechanisms responsible for the movements of the planets. Interestingly, Newton first had to create calculus in order to formulate the dynamic laws of motion. In his equations, Newton included the speed of planets in their orbits, their size and mass as measured over a unit of time.

Newton's equations enabled him to accurately predict the movements of the planets and stars in the universe. That key word, *prediction,* is the hallmark of science. If an observation is repeatable and predictable, it does not represent an accident or a chance event. While we may not see or understand the mechanism that produces a specific observation, the quality of prediction reveals there are defined rules governing that process.

As a young developmental biologist in 1967, I had the unique opportunity of cloning multipotential stem cells. I placed a single stem cell in a culture dish, and that cell divided every ten hours. After a week of cell doublings, the culture dish contained about thirty thousand cells, all of which are genetically identical because they were derived from a single parent cell. I divided the cell population into thirds and inoculated them into three culture dishes. The cultured cells' environment contained a specific growth medium, a solution that contains the substances required for growth; it's the laboratory version of blood.

Since I synthesized the culture medium in the lab, I was able to create three different versions of media with slight differences in their chemistry. Each dish of cells was fed a different version of culture media, a different environment. In one dish the cells formed muscle tissue, in the second dish they formed bone tissue, and in the third dish they formed fat cells. The genetically identical cells developed in three different expressions based on the chemistry of their environment. The study revealed that it was environmental signals that controlled gene activity. These observations contradicted the reigning central dogma of molecular biology, the belief that DNA is self-actualizing, being able to turn itself on and off in controlling biological life. My experiments showed that genes did not control their own expression; genes were being controlled by information from the environment.

At the time, my fellow biologists dismissed my observations because they were imbued with the conventional belief that genes control biology. My peers argued that my results were just "artifacts and that, for some unknown reason, the outcome was just an anomaly. In spite of their skepticism, my experiments could be replicated, and the results of the outcome could be predicted even before the experiment was initiated. Much like Newton could predict the movements of planets, I was able to predict the results in my cell cultures.

For more than twenty years, conventional researchers ignored my research. However, in 1990, science officially recognized the role of environment in shaping genetic activity when they established a new field of science, epigenetics. By then it was recognized that environment not only controlled gene activity but also that environmental signals could modify the readout of genes, without altering the original genetic code.

I am recounting my personal story because, today, Yury Kronn stands at the same point where I stood in 1967. As this book illustrates in detail, Kronn has done many scientific experiments with subtle energy that can be replicated with predictable results. While his work has passed the test of that essential trait of science—prediction—as of yet, there is no understanding of the mechanisms of subtle energy. While its impact is clearly observable, subtle energy cannot yet be measured and quantified. However, Yury's experiments are predictable, and by consequence are not related to chance. Clearly an unseen mechanism is being expressed in his experiments.

Science has a history of ignoring or downplaying new ideas. The lack of scientific interest in response to Kronn's research reveals that researchers are not interested in pursuing new ideas that challenge conventional thinking. The Newtonian concept of a universe containing two noninteracting realms of matter and energy still dominates the current biological paradigm. Based on that awareness, conventional science offers no consideration for immaterial, invisible forces influencing biology.

The separation of matter and energy realms was an important ploy by Newton. He purposely separated the material realm from the invisible energy, for the latter was referred to as spirituality, and that was the exclusive domain of the church, the power that controlled Western civilization at that time. This division of matter and energy allowed science to operate outside of the control of the church.

The focus of science on the material realm has profoundly influenced our current worldview. For instance, in adhering to the notion that matter can only be affected by matter, and not be influenced by energy, science focuses only on physical solutions to influence biology and its behavior. This is the power behind pharmaceutical companies whose sole mission is to make chemical (material) pills to adjust health and cure diseases: The pharmaceutical company is not interested in the current movement toward energy healing, simply because they cannot monetize energy in the form of a material pill.

In the past century, the foundations of science have begun to shake. First, it was discovered that the atom was not the smallest particle but instead consisted of even smaller invisible particles: protons, neutrons, and electrons. Then came quantum physics, the most tested and verified of all of the sciences. This new physics, now almost a hundred years old, has yet to be fully adopted by biomedical science.

Quantum physics has proved that there are particles even smaller than electrons, neutrons, and protons. But more importantly, according to quantum physics, these particles are actually made of energy vortices and are not expressed as matter. The message of quantum physics is that matter is an illusion. As Albert Einstein wrote, "Reality is an illusion, albeit a very persistent one." Quantum physics reveals that there is only one realm in the universe, energy, and there is no duality. Since energy has no borders, by definition that means that everything in the universe is entangled as energy fields. Quantum physics offers a very different understanding of the nature of life, the planet, and the universe.

Physics has instruments that can read the energy making up the electromagnetic spectrum, which accounts for only about 4 percent of the total identifiable mass-energy of the universe. The remaining 96 percent of our reality is conventionally postulated to be dark matter, or dark energy, which cannot be seen, nor are there any instruments to measure it. However, quoting Einstein one more time, "the invisible energy field is the sole governing agency of the particle (matter)." The point being that invisible forces are shaping our physical world. Yury Kronn and others show experiments that are influenced by an unrecognized energy realm they refer to as "subtle energy."

The uncontestable and absolutely valid truth is that invisible energy shapes our lives. The problem comes down to the fact that mainstream science is not considering the role of energy outside of the influence of the electromagnetic spectrum. The famous inventor and engineer Nikola Tesla said, "The day science begins to study non-physical phenomena, it will make more progress in one decade than in all the previous centuries of its existence."

This is where we stand today, and why Yury Kronn's book is of such tremendous value. In this book, Kronn describes his observations about scientific experiments with subtle energy. His experiments follow accepted scientific methods. However, just as happened to me more than fifty years ago, the scientific community does not accept Yury's observations without a clear understanding of, or ability to measure, subtle energy.

While science cannot explain the underlying mechanism accounting for Kronn's experiments, there must be an organizing factor or mechanism since his results are not random. Kronn's experiments can be predicted and replicated and, as such, are relevant in understanding the nature of life.

It is time for science to open up to assessing a bigger vision of the universe. This is of critical importance at this time, since humans have precipitated the sixth mass extinction event, which is now threatening all of life on the planet. It is vital to our survival for us to understand the influence of invisible forces. For us to survive, we need to become more fully aware of the mechanisms by which the universe operates and recognize the interconnectedness and harmony that permeates it. We cannot survive by continuing to follow the mission of science established in the 1600s by Francis Bacon, to control and dominate nature. The reason is we are not outside *observers* of nature, we *are* nature.

The experiments presented in this book undeniably confirm that subtle energy provides an additional energetic mechanism of epigenetics. That opens new ways of regulating the environmental influence on the biology of living organisms. This includes opportunities to compensate for the negative effects of energetic environmental pollution created by modern technologies.

In this book, Kronn describes that a negative attitude of a person can cancel the outcome of an experiment, whereas a person with a positive attitude can manifest a specific result in the same experiment. Such observations lead to the role of perhaps the most incomprehensible energy in our universe: consciousness. Consciousness is an energy that does not dissipate over time and distance. While we are aware that electromagnetic waves become weaker over time and space, the energy in Kronn's experiments is not affected by distance and time. Consciousness is not an electromagnetic field.

Nobody knew that there were quarks inside atoms until we built very expensive instruments that allowed us to observe them. Currently, science does not have instruments to measure subtle energy. But it is here, it is present, and we can use subtle energy to heal our planet—even if we do not yet recognize its mechanism. The observations in Kronn's research demand science to focus its attention on the role of subtle energies in shaping our world. When the mainstream scientific community is ready, the experimental methodology of subtle energy research presented in this book can be used as a starting point for the quantum leap of science predicted by Tesla. Our lives may just depend on that breakthrough.

BRUCE LIPTON, PH.D., is an internationally renowned developmental biologist whose pioneering research gave rise to the field of epigenetics. Recipient of the 2009 Goi Peace Award, he is the bestselling author of *The Biology of Belief, Spontaneous Evolution,* and *The Honeymoon Effect.*

INTRODUCTION

Investigating the Unexplainable

THERE IS A STORY that is frequently told at conferences or cited in books to illustrate how hard it is for humans to accept new concepts or ideas, or, conversely, how good we are at recycling existing perceptions without truly opening our eyes. The story goes like this: When Christopher Columbus first arrived at the shores of the new world, the Native Americans could not see his ships. The ships were in plain sight, but supposedly the Indians, as Columbus would misname them, did not have the proper mental slots or receptors to process or accept something they had never seen before.

The only factual evidence for the story does not come from the journals or the ship logs of Columbus and his crew but from an account written almost three centuries later by Joseph Banks, a botanist accompanying James Cook on his voyage exploring the east coast of Australia in 1770. Other parts of Australia had been discovered by Dutch and Spanish explorers in the 1600s, but Cook was the first to visit the eastern part of the continent. Banks writes about the Aboriginal fishermen they encountered: "[T]he ship passd within a quarter of a mile of them and yet they scarce lifted their eyes from their employment; I was almost inclind to think that attentive to their business and deafned by the noise of the surf they neither saw nor heard her go past them. . . . Not one was once observed to stop and look towards the ship; they pursued their way in all appearance entirely unmoved by the neighbourhood of so remarkable an object as a ship must necessarily be to people who have never seen one."

We tend to have sympathy for what we think of as these less developed cultures who perhaps lack the curiosity to observe the advance of technology. However, oftentimes in our modern world, all of us—scientists, thinkers, citizens—are closed into a certain way of seeing or thinking about things. We will not see what is directly in front of us. It is not that we cannot see. It is that we do not want to know. We are too busy fishing.

In the past almost half a century, I have seen many unexplainable ships sailing along coasts. I have witnessed phenomena for which science—at least as we know it—does not even have the beginning of an explanation. At the same time, today, there is a consensus that science can explain only 4 percent of our reality. Yes, that is all that Western science can explain after the proud discoveries by scientists of the past four hundred years, from René Descartes and Isaac Newton to Albert Einstein. According to the latest calculations, only 4 percent of the mass-energy of the universe is occupied by atomic matter and electromagnetic energy; that is, by stuff that we can see or measure. That leaves 96 percent, vaguely described as dark matter and dark energy, that we cannot see, do not understand, and cannot explain.

That 96 percent is here, everywhere and at every moment of our existence. Is it even realistic to assume that it has no influence on our lives? Can we afford to continue fishing as if nothing has changed now that we officially know that we can only explain such a tiny part of our reality? It is not that there are no ships sailing by, we are simply not conscious of them. We do not even understand the source of our consciousness. Even if we can explain the big bang, where we think this all started, we have no idea what caused that giant explosion so many millions of years ago. And so on. There is much to be discovered and explained.

This is a book about my journey of discovery. My journey is only the beginning of a much longer journey—after all, we still have to explore the mysteries covering 96 percent of our existence. This book documents my experiments, conducted as much as possible according to the rigor of today's science. These experiments need to be replicated and extended. I do not know whether there is an end to this journey. I do not know whether we will ever be able to fully explain 100 percent of our existence. However, I *do* know that this journey of discovery has major implications for the future health of the Earth and of her human inhabitants. Ecosystems and lives will be much healthier and more resilient if we better understand the universe in which we are living. Our world will be a much better world if we keep our eyes open and remain curious.

But first, let me tell you a bit about myself. My life began in a deeply polarized world. I was born in 1935 in Tashkent, which at that time was the most populated city in the Asian part of the Soviet Union; today it is the capital of Uzbekistan. Tashkent's history, as a major destination on the ancient Silk Road, stretches more than two thousand years and includes destruction by Genghis Khan in 1219. When I was a small boy, before World War II, my parents got divorced and I moved with my mother to Crimea on the northern coast of the Black Sea. When the war started, we were evacuated to a small town close to the Ural Mountains. Our life there was cold and hungry, but we survived. After the war we moved several times, and we ended up in Kuybyshev. Today that city, at the confluence of the Volga and Samara Rivers, is called Samara. It is one of the major cities of modern Russia.

I went to high school in Kuybyshev. The post-war world had become completely disconnected and divided. Our life was simple in the Cold War era. We were taught that what we did in the Soviet Union was right, and what people in the Western world did was wrong. The world was divided into black and white. There were no connections. We shared very little with the rest of the world. Of course we had our doubts, if not criticisms, but few questions were asked. Asking questions was sometimes dangerous. Once, during a lesson on modern (which meant Soviet) history, I expressed my curiosity at how Joseph Stalin, who at that time was seventy-two years old, was able to simultaneously fulfill both main political positions in the Soviet Union: General Secretary of the Communist Party and Premier of the Soviet state. The teacher immediately stopped the lesson; I was taken to the office of the school's director, and my mother was summoned there. The director locked the door of his office and he and the history teacher, both pale and trembling from fear, questioned me and my mother about how I could have asked such a terrible question. Finally, after I had answered repeatedly that nobody else was involved and that it was just my own spontaneous question, we were allowed to go home with the strict order not to discuss the event with anybody. At home, my crying mother explained to me that both of us, as well as the school director and the teacher, could be imprisoned if the event would become known to authorities. We were lucky that did not happen.

I graduated with a silver medal. I did not get a gold medal because I received a B for Russian. I was told I was not expressive in my writing about Soviet life. In reality, I struggled writing about all the communist propaganda. However, I made no grammar mistakes, and they could not give me less than a B. Social sciences and politics were controversial subjects in the Soviet Union. That is why I, and most of my friends, were interested in science: chemistry, mathematics, physics, biology, and so on. I was attracted to physics because it is the science of the construction and the laws of the universe.

I wanted to go to the Moscow Institute of Physics and Technology, a prestigious institute related to Moscow State University. However, I was not accepted. "How come? I only have one B," I asked the admission committee. "We do not need to explain that to you. Please take your documents and go home," they replied. Afterward, I visited my father in Moscow. He looked over my application materials and immediately shouted, "What did you do? Are you crazy?"

Applying for university in the Soviet Union meant filling out many papers. On one, I was asked where my parents and grandparents were born, and I had written down that my father was born in Greece.

"How could you do that?" my father asked.

"But it is true," I retorted.

"It does not matter what is true. What matters is how it should be. I was born in the Donbass region of Russia."

"How come?" I asked my father.

"Very simple. Old documents were lost during World War II. I said that I was born in Donbass and I was registered as a Russian citizen."

"How can you be Russian?" I asked him. "Your father was Greek and your mother Jewish."

"That is why I am Russian," my father replied, laughing.

I re-did my application forms including my "Russian father" and applied for the State University of Gorky, which, I was told, had a secret faculty for modern physics. I was admitted only to discover, to my great regret, that there was no secret physics program. I was accepted into the chemistry department, but I did not want to study chemistry. I was lucky that Gorky also had a radiophysics faculty set up by Aleksandr Andronov, an academic who pioneered a new approach to applying mathematics to physics. I managed to get transferred to this very good school, from which I graduated in 1958. Only later did I realize that this course of events had brought me to exactly the right place for my education. At that time, all universities in the Soviet Union were focused on teaching specific information. They were not teaching thinking, as that could only generate trouble. However, Andronov's radiophysics faculty at Gorky was a rare place where students were taught how to *think* about physical phenomena.

I learned to think, but, in the Soviet Union, that did not mean I could choose my own job. University graduates were distributed, and I was sent to a kind of military design bureau that supplied the army with electronics for rockets. I never saw a rocket, but I invented a new generator for the rocket's radar. The director of the Gorky Radiophysics Institute, Maria Grechova, had noted my work, and she invited me to become her postgraduate student. She convinced me to focus my Ph.D. on nonlinear optics, the science of the interaction of laser radiation with matter, a new field that was fast emerging at that time. I developed and published equations describing the nonlinear interactions that happen when atoms and molecules are hit by powerful light. This work became the basis of my Ph.D. dissertation.

After my postgraduate studies, I moved to one of the scientific towns created by Nikita Khrushchev in the Moscow region, and I was given a position as senior researcher at the Institute of Radio-engineering and Electronics of the Academy of Sciences. One day, I presented my research at the department of Wave Processes at Moscow University, headed by academic Rem Vicktorovich Khokhlov. Khokhlov praised my work and suggested that I become a theoretician-supervisor of a group of experimenters that he had put together to work on elucidation of some nonlinear optic phenomena. The move to Khokhlov's department at Moscow University would dramatically change my career.

Khokhlov was a remarkable man. He was a respected physicist, but he was also popular in the Central Committee of the Communist Party. He was close to Mikhail Suslov, the second-most powerful man in the Soviet Union after General Secretary Leonid Brezhnev. Because of that connection, Khokhlov had considerable freedom in choosing the direction of his scientific and organizational pursuits inside the Soviet Academy of Sciences.

I worked with Khokhlov and his team of experimenters for twelve years. During that time, I gained invaluable experience in planning experiments aimed at explaining incomprehensible phenomena. We illuminated many wonders of nonlinear optics that had been puzzling scientists for many years. We published many articles about our findings in the leading Soviet scientific journals. And we were the first in the world to create a tunable source of laser radiation in the infrared area of the light spectrum.

Khokhlov kept a close personal interest in our research, and over the years our regular meetings grew into a dear friendship. In the beginning of 1976, I reported to him that our group had completed all tasks formulated by him ten years prior. Khokhlov was very satisfied, and after I finished my detailed report, he put his hands on my shoulders and said that my work was worth a doctorate degree. Such a degree—a step up from the regular Ph.D. in the Soviet academic system—was intended for "scientists who opened a new direction in their branch of science." Khokhlov suggested that I take a half-year sabbatical to prepare for the defense of my doctorate dissertation at the science council of Lebedev Physical Institute of Moscow (FIAN), headed by Nobel Prize–winner Nikolay Basov.

Before I tell you about my experience at the FIAN science council, I need to take you back to some strange and unexpected events that had happened in my life. I had encountered a group of scientists who were regularly discussing paranormal phenomena. Being curious and open-minded, I joined this group. The group was brought together by Sergei Mitrofanov, an engineer who later graduated from a medical school and became a medical doctor. He was born with psychic abilities, and he thought that it would change civilization if he could bring a group of scientists together to study these phenomena. His initiative attracted a diverse group of scientists. There were biologists, mathematicians, and physicists like me. Sergei Mitrofanov taught us meditation and trained us to sense subtle energy. I discovered that I am able to feel something around human bodies.

Mitrofanov told us about the chakras, the energy centers of the body. According to ancient ayurvedic medicine, certain mantras stimulate certain chakras, and we did a scientific experiment to research this correlation. One person was put in a room with a box containing mantras written on pieces of paper. The person randomly chose one and began repeating the mantra for fifteen minutes. Subsequently, the other members of the group entered the room one by one to try to feel which chakra of the person reading the mantra had been activated. We repeated this research for a while. Our results showed that in more than 90 percent of the cases, the group members were able to read the chakra that, according to ayurveda, relates to the chosen mantra.

I realized the science I had been studying was missing something very important. Soon, I experienced an even stronger confirmation of that insight. A colleague at the Institute of Radio-engineering and Electronics told me about secret experiments that were being done at our institute with two women with paranormal capabilities. In one experiment, ultrasound signals were propagated through a large man-made ruby crystal. One of the women, Ninel Kulagina, was able to stop the ultrasound by simply putting her hands around the crystal—the signals on the monitor disappeared. We thought that the generator failed, but when she took her hand off the crystal, the signal came back. She repeated the same thing three times. After that, she looked very pale and was extremely tired. I wondered what kind of energy she used to cause the effect. How could her energy stop propagation of ultrasound signals in the world's second-hardest material?

In another experiment that was replicated multiple times, the second woman, Rosa Kuleshova, was able to read random pages of a book hidden under a cover, even when it was a metallic plate. With my own eyes, I saw things that had no place in my physicist's mind. The experiences were completely unexplainable from the perspective of physics. I was shocked. I remember that I could not sleep the night after I first witnessed these experiments. "This is not logical, impossible," I said to myself. "I'm a scientist." Ultimately, I needed to admit that what I saw with my own eyes were facts.

These facts set the stage for a mystical experience I had right after my meeting with Khokhlov, when he proposed that I qualified for a doctoral dissertation. Looking back today, that experience led me to my life's mission and to the work that I will describe in this book. The evening I came home from my meeting with Khokhlov, I was very happy. With his academic support, my career seemed forged. That night when I was about to fall asleep, all of a sudden, I heard a clear voice in my head: "This is not what you are supposed to do." I did not pay much attention to it. What a stupid thought, I told myself, and went to sleep. The next morning when I woke up, I heard the same voice again: "This is not what you are supposed to do." This time I said, "What am I supposed to do?" "You will find out soon," the voice responded. I had never had any similar strange experiences. I did not know what to make of it, and I continued with my dissertation.

The final presentation of my dissertation at the science council of FIAN was scheduled for June 7, 1977. A week before that date, the secretary of Khokhlov, who had become rector of Moscow University in 1973, reached out to me and told me that the rector urgently wanted to see me. We met at the end of the workday and we were alone in his office. He congratulated me on accomplishing my dissertation and asked what my next plans were. I told him I was interested in investigating biological applications of lasers. "Very interesting," he said, "but what about the experiments with Kulagina and Kuleshova that are going on at your institute?"

I was stunned.

"Do you know about that?" I asked, surprised.

"Yes, I have met with Ninel Kulagina. She held her hand half a meter from my head, and I felt she was pressing my skull. This energy is real. We need to research it."

"I am already trying to do this," I said.

Khokhlov was very happy to hear that. He told me that he was destined to become president of the Academy of Sciences that autumn, and he was planning to open an institute of nonlinear optics and was going to appoint me as head of the lab at the new institute. However, my appointment was meant to be a disguise. "The real research you are going to do will be into these strange phenomena, but nobody will know about it," Khokhlov said.

He asked whether I had some fellow scientists I could invite to participate in the new lab. I told him that I could bring some open-minded members of Sergei Mitrofanov's group.

"That is good. Pack your suitcases and get prepared. When I come back from summer vacation, we will start."

My doctoral dissertation, "Resonant Non-Linear Interactions of Light with Matter," was unanimously accepted by the science council of FIAN. Rem Khokhlov was a member of the council, and that day I saw him for the last time.

He went to climb Communism Peak, now called Ismoil Somoni Peak, in the Russian part of Tibet. While he was climbing the mountain, he suffered a heart attack. He was rescued with a helicopter, and doctors said he would recover in a few weeks. But then strange things began to happen. A special aircraft came from Moscow with a medical team with strict orders to take Khokhlov to the Kremlin hospital. Nobody except his wife was allowed to visit him there. She was concerned and went to her contacts at the Central Committee. They called the hospital, and the staff assured them that Khokhlov was kept in a quiet environment for his own best interests for a few days and that he would be able to receive visitors after the following weekend. That Saturday, a commission of doctors decided he needed a blood transfusion, even though there was no indication that recovering from a heart attack required a blood transfusion. The blood transfusion instantly killed Khokhlov.

It was never officially accepted, but—as events were pieced together afterward—it is clear that Khokhlov was killed by the KGB. We do not know why. We know that the KGB was doing research into psychic phenomena. They definitely knew that Khokhlov was about to launch an extraordinary scientific research project into these same phenomena, and they likely did not want him to do that outside of their control.

The events around my meeting with Khokhlov and his sudden death confirmed my own mystical experience: I knew what I was supposed to do. We continued to do some research with the group around Sergei Mitrofanov. We began investigating prayers and tools used in rituals of different religions. We found that some of them, like the tefillin that is worn by observant adult Jews during weekday morning prayers, strongly enhances energy flow through the body. We researched the effect of mantras with the Hare Krishna movement that appeared in Russia at that time. One of these meetings was discovered by the KGB. We were all arrested, and, subsequently, I was dismissed as head of the lab. When I asked the party committee for an explanation, I was told that what we were doing was religious propaganda and that, as a Soviet scientist, I should explain the communist ideology to people.

It was 1982. My frustration with Soviet reality grew, and I decided to join a dissident movement. With ten other people, I began organizing what became the Moscow Trust Group. Our objective was to foster trust and promote peace between the Soviet Union and the United States. In an appeal to the governments of both superpowers, we wrote: "The USSR and the USA have the means to kill in such proportions that would end the history of mankind. A balance of terror cannot be a reliable guarantee of safety in the world. Only trust between peoples can create a firm assurance of the future."

Starting the Trust Group did result in losing my job completely. I kept a tiny income because friends were able to give me a job as a watchman of dachas, the vacation homes of the Moscow elite, during the long winter when nobody was there. The Trust Group was very active. We had regular meetings at the American Embassy. We published articles and gave interviews. Of course, our activities were closely monitored. Together with another member of the Trust Group, Professor Yuri Medvedkov, I was put in prison for two weeks, and in England a special demonstration was organized in our defense. I was interrogated by the KGB many, many times. But they could not suppress us as we only spoke about the need for friendship and trust; we did not even promote disarmament.

In 1987, when Mikhail Gorbachev had launched his perestroika, he released many political prisoners. Several members of the Trust Group joined the Glasnost Press Club headed by prominent writer Lev Timofeev, and together we organized the first Moscow International Symposium for Humanitarian Problems. I chaired the disarmament section of the symposium. Four hundred people came from all over the Soviet Union. We came together in private apartments as we were not allowed to use official buildings. The event was covered by some forty foreign reporters and broadcasted all over the world. After the event, I was called to the immigration office, and I was told that I had "received an invitation to go to Israel." I never saw the invitation, but in April 1988, I left for Vienna. There I was given a choice to go to Israel or to the United States. I chose the United States, and, after waiting for three months for a visa in a refugee camp near Rome, I came to New York.

I needed to find work. My scientific credentials brought me in touch with Columbia University, and I was offered a job as a research scientist. However, my interest in investigating unexplainable phenomena had not faded. During my first months in New York, I was introduced to a small company from Illinois that claimed to be working on some kind of energy medicine technology. I thought they were serious people. As I was so eager to do the research on subtle energy, I made the rather irresponsible decision to turn down the offer of Columbia University, and instead I joined that company. I soon discovered that their equipment was totally inappropriate and that they had only hired me to advertise that they were working with a Russian scientist. Our collaboration was short-lived, but the experience put me on the right track. During this time, I built my first generator of what we today call "subtle energy," and I developed a method to generate and record specific energy patterns—for meditation, sleep, concentration, and the like. My first recording device was a reel-to-reel tape recorder. I created sets of magnetic tapes playing soundless subtle energy "music" through a regular tape player.

In the autumn of 1990, I left the company in Illinois and moved to Columbus, Ohio, to start my own business. With the help of psychic people, I continued to develop energy patterns for specific experiences. I developed a set of sixteen vital energy tapes that today, two decades later, people are still using. The research and experiments that I will describe in this book are all based on this initial work that was inspired by my early experiences in the Soviet Union.

In 2000, together with my wife, Constance, I started my present company, Energy Tools International. In 2004, we moved to southern Oregon. That is where my journey as a physicist, from the tight framework of Soviet materialistic ideology to the frontiers of metaphysical science, has led me. In the past fifteen years, I have been working in this pristine and peaceful environment, which is essential for the research and experiments I am conducting.

About the Book

In the first chapter of this book, I will discuss the nature of subtle energy and where this energy fits in the universe. Here, I will argue that subtle energy should not be interpreted as a by-product of the forces known to science. We will contemplate developing a methodology of subtle energy research.

In chapter 2, we will review experimental facts illustrating that subtle energy belongs to the subatomic world and follows laws that are fundamentally different from those governing the energies known to our science. Here, I argue that the substance of a subatomic, or dark matter, world participates in the human structure—as the "subtle body," as described by the ancients.

Next, in chapter 3, we will look into the possible mechanisms of subtle energy's interaction with physical matter and with the human body. This discourse is based on the fascinating extrasensory observations of quarks and superstrings by Annie Besant and Charles Leadbeater and the scientific analysis and confirmation of their work by nuclear physicist Stephen M. Phillips, Ph.D. We will see how research conducted with people with extrasensory perception, or human instruments, allows for establishing a hypothesis about the properties of subtle energy patterns of physical substances.

Chapter 4 describes the Vital Force Technology (VFT) I developed to harness subtle energy. This technology makes it possible to satisfy demands of modern scientific methodology for stability and repeatability in experimental processes. It opens the door for further rigorous scientific research on the properties and potential applications of subtle energy. Several radically different methods of creating subtle energy patterns will be presented. We will see how bridging ancient knowledge and modern technology allows for creating sophisticated formulations of subtle energy for a variety of practical applications.

Chapter 5 presents experimental results of several energy formulas' influence on plant germination and growth. From basic living organisms, we will move on to experiments with animals and people in chapter 6. We will discover experimental proof that the mind interacts with matter by means of subtle energy, programming the subtle energy *to do the job.* This exploration gives us a key to understanding the mechanisms of the placebo effect, as well as the potential healing power of affirmations, prayers, and the spectrum of "biology of belief" phenomena. We will explore the possible participation of subtle energy in epigenetics, the altering of gene function without altering DNA.

In chapter 7, I demonstrate the possibility of using subtle energy for creating clean and energetic pollution-free environments for vitality and better healing. In the concluding chapter, I present opportunities for further research and development to broaden and improve our understanding of the 96 percent of the world that is around us yet still not understood. That is still a giant task that requires bold ambition.

In chapters 5, 6, and 7, I describe the experiments we conducted in language for general readers. There are appendices for detailed descriptions and analyses, too, which will allow scientists and researchers to replicate and build on the experiments. With this book, the work presented here, and the work that will continue, we are entering a field of science like we have never seen before. "Talent hits a target no one else can hit; Genius hits a target no one else can see," said German philosopher Arthur Schopenhauer. We need many open-minded "genius scientists" to investigate phenomena like the ones presented in this book, which we can observe but not yet fully explain. This is a field of science that challenges core values—like objectivity—of the current scientific paradigm. In the world of subtle energy, like in the world of quantum mechanics, the observer is always also a participant. When you try to record the healing natural energy of a waterfall while you are thinking about the quarrel you just had with your wife, you will be recording the energy created during the quarrel with your wife as well. Your own energy will distort the energy pattern you are recording. In this book, we are describing phenomena and experiments that interact with individual consciousness. That is why the emerging science of subtle energy challenges everything we know. It is *so* not simple.

However, that unknown and that complexity come with tremendous opportunities for healing ourselves and the planet—for healthier life on all levels—from physiology to spirituality. Welcome to the 96 percent!

PART 1

The Powers That Surround Us

1

A Force We All Experience

"Use the Force, Luke!"

EVERYONE WHO HEARS that phrase instinctively knows what Luke Skywalker's mentor Obi-Wan Kenobi is talking about in *Star Wars.* We intuitively grasp that such a force exists. It's palpable to us, at times, when we enter a room and immediately sense a buoyant, uplifting atmosphere, even though we don't know the cause. Suddenly our vision becomes clearer and our mood better. Conversely, we all have walked into a room in which there's a heavy vibration in the air as though some unpleasant force is present. Sometimes, people we've never met ignite our interest and sympathy from the first minute. We simply feel comfortable with them. With others, we may feel uncomfortable for no apparent reason and want to leave their presence as soon as we can. Typically, we say we just do not like the "energy" we are picking up from them.

"It's an energy field created by all living things. It surrounds us, penetrates us . . . and it binds the galaxy together," says Obi-Wan Kenobi to Luke Skywalker. The description and principles of the *Star Wars* force resonate deeply with core concepts found in major spiritual traditions. Hinduism refers to *prana,* the vital, life-sustaining force present throughout the universe. In Buddhism and Taoism, the universal energy or life force that permeates all things, both living and inanimate, is called chi, ki, or qi. Indigenous traditions in many countries also mention and utilize a universal force for the sustenance of life.

While this force is instinctively known by billions of people and has been discussed in ancient scriptures and world literature for ages, it is nonetheless rarely, if ever, the focus of modern scientific study. There has been little scientific effort to determine whether such a force exists in any demonstrable, verifiable way. Scientists have hardly explored the nature and relationship of this force with the electromagnetic force and other forces known to science. Only few have studied the interaction of this force with matter and the human body. As a result, today, we have very little understanding about this force and its application or use to support daily life and human well-being.

Mainstream science dismisses the existence of the concept of the life force and openly ignores the knowledge collected by our ancestors through millennia of experience. Most scientists refuse to explore the obvious and potentially life-transforming questions that the references to the force point to. In *The Holographic Universe*, bestselling author Michael Talbot explains this lack of interest: "We are addicted to our beliefs and we do act like addicts when someone tries to wrest from us the powerful opium of our dogmas."[1]

Some have been brave enough to escape these dogmas and vested interests. Those researchers in frontier science—a reference that aptly describes their attempts to investigate *outside the box* of the official scientific paradigm—have conducted experiments on various properties of this subtle energy. The term points to the fact that this life force is not directly measurable by current scientific equipment.

Many books have been written about subtle energy. Most of them are devoted to its applications for healing. They discuss the science behind the principles of traditional Chinese medicine or examine energy medicine based on the use of modern technologies.[1–9] Other modern authors have explored the effect of the human mind on matter and on biological processes in living organisms, including the human body. These explorations inevitably arrive at one conclusion: Subtle energy participates in the communication between consciousness and the physical world in which we live.[1,6–11]

A growing stream of information indicates the emergence of a parallel science that is working on bridging the gap between the conscious universe of ancient thinkers and the modern scientific perspective of the universe as a mechanical machine. In its attempt to create a theory of everything, conventional science does not include consciousness. That is why Hungarian philosopher of science Ervin Laszlo, generally recognized as the founder of systems philosophy and general evolution theory, states: "At the most, physicists would come up with a physical [theory of everything]—with a theory that is not a theory of every *thing;* only of every *physical* thing. . . . There is more to the universe than vibrating strings and related quantum events. Life, mind, culture and consciousness are part of the world's reality and a genuine theory of everything would take them into account, as well."[12]

While modern science does not include the life force in its consideration of the forces shaping the universe, one of the giants of ancient philosophy, Lao-tzu, emphasized the general importance of it in his book *Hua Hu Ching* in 500 BCE! Lao-tzu argued that subtle energy connected universal consciousness and the material world and played a decisive role in human life[13]: "There are numberless energy rays in the Universe. . . . All rays . . . come from the subtle nature of the Universe. . . . The interwoven energy net influences the lives of individual human beings, whole societies and entire races. . . . A virtuous individual who responds to the high, pure, harmonious subtle energy rays and integrates them with the positive elements of his own inner being may strengthen his life, enhance his health and power, and lengthen his years."

Twenty-five hundred years ago, Lao-tzu and his contemporaries saw a complete and unified world. Today, we live in a polarized world where conventional science denies even the *existence* of the force that ancient wisdom presents as the *driving force of the phenomena of all life.* Is it possible that the action of this subtle force is so hidden that it is not easily seen using Western scientific methods? If yes, what mechanisms does this elusive force use in interactions with physical substance to be so invisible to scientific observation? To answer such questions, we need to review the facts that frontier science has discovered and collected about subtle energy.

One of the pioneers of the research of subtle energy is the MIT- and Princeton-educated physicist Claude Swanson, Ph.D. In his seven-hundred-page book, *Life Force, the Scientific Basis,*[14] he presents a wide range of experimental research on subtle energy compiled from all over the world. In the beginning of his book, Swanson illustrates the challenges for those who get involved in researching subtle energy.

> It manifests itself in many ways. Usually, a scientist examines only one aspect of it at a time. Therefore, most scientists who investigate it only see a part of the picture. Their description is reminiscent of the fable of the six blind men and the elephant. One man felt its tail and concluded it was a rope. Another felt its leg and described it as a tree. A third felt its ear, and described it as a large leaf, while a fourth touched its side and described it as a wall. In the case of subtle energy, it alters the other physical laws in many different ways. It responds to and affects consciousness, and also modifies electricity, magnetism, gravity, time, even nuclear processes. It is a many-faceted phenomenon. Yet in most cases, its effects are weak, so it keeps being discovered and forgotten over and over by many scientists throughout history.[14]

After reviewing an impressive list of physical phenomena affected by subtle energy, Dr. Swanson concludes that subtle energy is omnipresent and elusive at the same time.

> A scientific revolution is underway, as the effects of the "Life Force" are being demonstrated, documented and measured. It is like no other force known to science. It responds to consciousness and alters the other basic laws of physics. It is intimately involved with life processes and plays a central role in growth and healing. It does not weaken with distance and penetrates most materials and

> shielding. In many ancient cultures, manipulation of this energy is one of the central arts of medicine. It is present in many aspects of life, but in sufficiently small concentrations that it is usually overlooked.[14]

The last phrase of this quote raises an important question: What is a "sufficiently small concentration" of subtle energy? We don't currently have tools to make direct measurements of subtle energy. We can only observe and measure its effects on inanimate matter or on living organisms. This condition makes pinning down quantities and qualities of subtle energy both elusive and problematic, at best. However, we have an example of a dynamic, global phenomenon that is continuously supported and governed by subtle energy, and we observe it wherever we are on Earth. This phenomenon is us, human beings, and all other forms of life. "Our inner energy is simply a reflection of the greater play of energies in the universe as a whole," writes Richard Chin in *The Energy Within: The Science Behind Eastern Healing Techniques*.[2]

If we have an abundance of subtle energy everywhere on Earth to support all life (and we will eventually see there is no place in the universe lacking subtle energy), we don't need special conditions to observe its effects on physical phenomena and inanimate matter. Aside from the amount of subtle energy, other factors must be involved in the interactions between subtle energy and the physical world, defining whether a particular type of this interaction happens. Is it possible that subtle energy patterns—Lao-tzu's "energy rays"—are different in terms of the information encoded in them? An argument for this assumption can be found in the research on an entirely unique property of subtle energy: its ability to follow commands of the human mind. Some truly impressive effects of subtle energy on physical matter have been observed when human beings, specifically qi gong masters, manipulate this life force.

Chinese physicists have described their experiments with one well-known qi gong master: "Experiments with Dr. Yan Xin showed that a wide variety of physical and chemical processes can be affected. These include radioactive decay rate, the bonding characteristics of water, generation of infrared (IR) pulsed energy, creation of pulsed and DC magnetic fields, scintillation (light emission) of lithium fluoride crystals and the bending of laser beams. The experiments illustrate the wide spectrum of effects Qi (subtle energy) can produce."[19]

The question arises: What is the difference in all these experiments that leads to such different outcomes? The only difference was the intent of Dr. Xin. In each experiment, he intended to influence parameters of a different process. Does this mean his mind was programming the subtle energy to do what needed to be done? In other words, was the subtle energy in the different experiments encoded by the human mind with different information? It looks like it was, though we don't yet know how this mind/energy interaction works.

A skeptic would stop us here. How do we know any kind of energy was involved in this process? As we have said before, subtle energy is—not yet—directly measurable. Scientifically speaking, we can register only two events: the mental effort of the qi gong master and the resulting changes in the physical/chemical processes he intended to influence. These experiments don't prove that some energy *causes* changes—though it is a logical assumption. For mainstream science, it would be more persuasive if a qi gong master could be replaced by some kind of subtle energy-generating equipment, which would be capable of repeatedly programming subtle energy to produce the same effects as the energy generated by the qi gong master. If this were the case, the demands of modern scientific methodology for stability and repeatability in the experimental process would be satisfied. Additionally, this would theoretically supply proof to show that subtle energy interacts differently with physical matter, depending on the information encoded in the energy. As we will see in chapter 4, this type of electronic qi gong master already exists.

In the case of electromagnetic (EM) energy, we know it affects physical matter through its interaction with electrically charged particles. This interaction allows us to measure EM energy. We cannot measure subtle energy because *it does not interact with electrically charged particles.* Nevertheless, it interacts with physical matter and produces noticeable changes—ranging from healing effects in human cells to changing radioactive decay rates. This leads to a question of paramount importance, one that is raised by all the experimental facts collected by frontier science: What is the mechanism of interaction of subtle energy with physical matter? If we only look at the number of terms that are being used to describe subtle energy, such as orgone, bioplasma, scalar waves, torsion field, deltrons, to name a few,[14] we realize that the community of frontier scientists is not yet close to a consensus about the nature of subtle energy.

It is important to note that mainstream science can observe and measure the known energy fields, but it cannot answer the questions "What is the energy?" or "How is it made?" For example, we don't know *what* the electromagnetic field is or exactly *how* an electron is made. All we know is how electrical and magnetic fields interact with each other and with electrically charged particles. These interactions are described by the famous Maxwell equations, which were created on the basis of the fundamental experiments of Michael Faraday and others. Maxwell's equations precisely describe the behavior of the electromagnetic field in space and time, and its interaction with physical matter. They adequately predict all possible results of any experiments involving electromagnetism, and they've made it possible to invent many of the technologies we use every day.

In contrast, when we look at the various theories about subtle energy, we have quite a different picture. No unified theory exists, and different researchers suggest entirely different mechanisms for the production of subtle energy in nature. These suggestions vary widely, from considering subtle energy a longitudinal electrical or magnetic scalar wave[15] to "being produced by the interactions between elementary particle spins, which generate a twisting of space."[14] The trouble is that none of these theories explain the healing five elements qualities of subtle energy known and used for millennia in practical applications by acupuncturists and other traditional Chinese medicine practitioners. According to Eastern traditions, the elemental subtle energy properties of fire, water, earth, metal, and wood balance the energy flow in the human body and thereby maintain health and well-being. They allegorically emphasize the properties of different ranges of subtle energy and how they interact with each other. For instance, fire supports earth, earth supports metal, metal suppresses (or cuts) wood, fire suppresses (or melts) metal, and so on.*

The challenge of constructing an adequate theory of subtle energy may very well be connected to the difficulties mainstream scientists are facing when trying to create a comprehensive theory of subatomic particles. As scientists recently realized, subatomic particles originate in a multidimensional space. Thus, it has been theorized that what we observe in our three-dimensional physical world is some kind of projection of the events taking place in a multidimensional "metaverse," as it was called by Ervin Laszlo,[12] a state that is largely unimaginable to us. It is much easier for science to explain phenomena that happen in a world with which our senses are familiar. All our measuring devices, in one way or another, involve electromagnetic energy; therefore, we can say all of the equipment used by modern science represents an extension of our sensory perception, mainly our vision, which is based on the range of electromagnetic energy we call the visible light spectrum. Like our physical vision, all "extensions" of our vision act only in the three-dimensional world.

As soon as a phenomenon cannot be perceived by our physical senses and their extensions, all theories about the phenomenon resemble the theories of a blind man on "how to use colors for creating art." Here's a good example: Scientists are struggling to make sense of what they're calling "dark matter" and "dark energy," which occupies 96 percent of the universe (more about this in chapter 2). Subtle energy, according to both ancient Eastern philosophy and modern

*More on this subject can be found by reviewing sources 2, 14, 16, 17, 18, and 20 in this chapter's endnotes.

frontier science, definitely belongs to the same kind of phenomena as dark matter and dark energy: we cannot measure it directly, and we can only measure the effects it produces in our three-dimensional world. There must be a more effective way to study subtle energy than creating new "blind man" theories! After all, the practical reason to research subtle energy is to find a means to harness it for our needs—whether it is for our health and well-being or spiritual growth—*and* for finding nontoxic alternatives to existing technologies that are polluting our environment.

History shows us a different way to research and investigate subtle energy. Our ancestors knew that the forces of nature ruled the world. Whatever changes happened in their world, some force of nature was the cause. For example, thousands of years ago, the Assyrians knew that water flowed downhill and that a force of nature was responsible. They built aqueducts—sophisticated irrigation systems—to carry water efficiently from one place to another, using this force (which only many centuries later would be called gravity), and practically applied their knowledge of this force to build monumental structures that have lasted thousands of years . . . all *without* the use of mathematics.

Our ancestors did whatever they could to harness the forces of nature. It did not bother them that some of these forces were invisible, nor that they lacked understanding of the precise laws governing them. They were observing the effects produced by these forces and made attempts to use them for practical purposes. In fact, whole civilizations arose and functioned using these forces. And sometimes they achieved goals that are beyond even the capabilities of today's technologies! As you look at this twelve-hundred-ton stone (see fig. 1.1), observe how precisely it was cut and imagine how it was moved. Today, we can replicate these achievements with modern equipment that only very recently has become available. Eastern civilizations long ago developed very sophisticated systems to employ the energies of nature, deal with the environment, and support peoples' well-being, using forces that modern science still does not know about.

Fig. 1.1. The Stone of the South, a Roman monolith found in Lebanon, is one of the largest ever quarried. Ancient civilizations used invisible "forces of nature" for practical purposes, long before modern machinery existed.

We can say that, in the distant past the methods used for investigating the universe were phenomenological: researchers observed various phenomena and made attempts to develop practical models and describe rules that made it possible to use these phenomena. One of these practical models is the principle of the five elements. Another famous model of universal energies functioning in the world is the Tree of Life (used in the esoteric tradition of the Kabbalah).

Speaking in modern scientific language, these ancient researchers developed algorithms of natural phenomena, which of course included energies ("emanations," as they were described in the Tree of Life model). As Ervin Laszlo writes, "It is enough to identify the basic constituents of a system and give the rules—the algorithms—that govern their behavior. A finite and surprisingly simple set of basic elements governed by a small set of algorithms can generate great and seemingly incomprehensible complexity merely by allowing the process to unfold in time. A set of rules informing a set of elements initiates a process that orders and organizes the elements, so that they create more and more complex structures and interrelations."[14]

Our goal is to present a "set of basic elements governed by a small set of algorithms" that can be applied to subtle energy for the purpose of harnessing it and making it accessible for rigorous scientific research and beneficial, practical applications. We will do this through an analysis of the results of crucial experiments conducted by mainstream and frontier scientists, including the author's thirty years of research into the nature and fundamental properties of subtle energy. We invite you to participate in this analysis, applying reason and common sense to the process.

2

The Nature of Subtle Energy

"But I'm *not a serpent, I tell you!" said Alice. "I'm a—*
. . . I'm a— . . ."
"Well! What are you?" said the Pigeon.

ALICE'S ADVENTURES IN WONDERLAND
BY LEWIS CARROLL

CURRENT SCIENTIFIC THEORY holds that all substances of our physical and chemical world are made from the chemical elements listed in the periodic table of elements, first formulated by Russian chemist Dmitri Mendeleev in 1869. The modern periodic table starts from the lightest element—hydrogen—and advances through progressively heavier and heavier elements, like lithium, carbon, magnesium, calcium, chromium, iron, silver, and gold, then further to radioactive uranium, finishing with the elements like copernicium (atomic number 112), and so on, which are artificially created by nuclear bombardment and very short-lived. As we examine the influence that subtle energy has on numerous substances, it is important to note and remember that all elements, and thus all substances in the billions of galaxies in our universe, are combinations of only *three* stable elementary particles: the proton, neutron, and electron. If subtle energy can be seen to affect these subatomic particles, then it has potential to play a role in everything in the universe. The nucleus of all elements is made up of various combinations of protons and neutrons; the former have a positive electrical

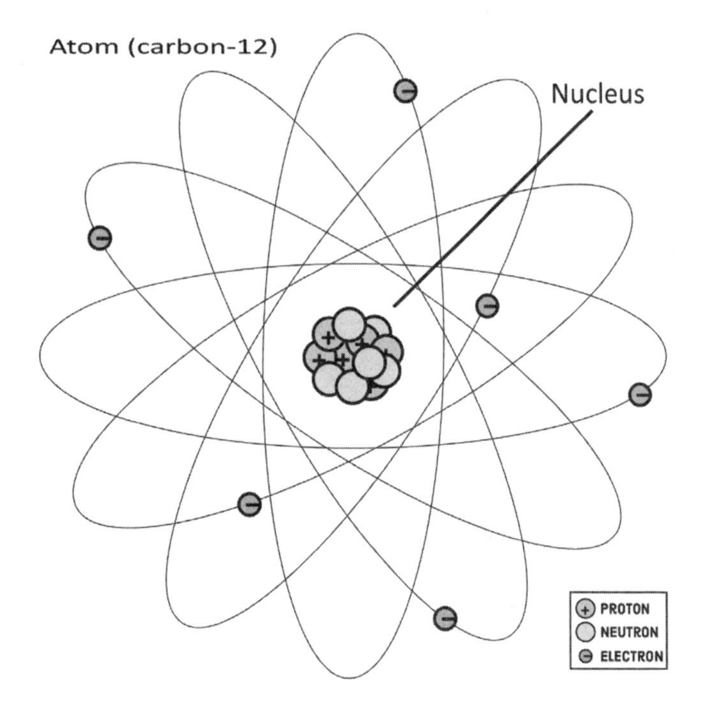

Fig. 2.1. A diagram of a neutral carbon-12 atom. Carbon is balanced with six protons, six neutrons, and six electrons.

charge, and the latter are neutral. The number of electrons on the orbits around the nuclei of each atomic element is equal to the number of protons, which makes atoms electrically neutral. Different atoms have different numbers of electrons on their external orbits, and this number defines the difference in their chemical properties.

Following decades of theories and very costly experiments using particle accelerators, scientists have come to the conclusion that the proton and neutron are not elementary particles but instead contain much smaller objects, which they named quarks. Among the six quarks known to science, only two of them—the "up" and "down" quarks—participate in the creation of the proton and neutron. These quarks have fractional electric charge values: the up quark has $+2/3$ of the value of the electron's charge, and the down quark has $-1/3$ of that value.

Combining in sets of three, the up and down quarks create protons and neutrons:

2 up + 1 down = proton, with an electrical charge of +1 => $(+2 \times \frac{2}{3}) - \frac{1}{3} = +1$

1 up + 2 down = neutron, with an electrical charge of 0 => $\frac{2}{3} + (2 \times -\frac{1}{3}) = 0$

Some physicists have suggested that quarks are not the most fundamental particles but are composed of three subquarks.* As it stands today, science does not know whether this is true. So far, the ladder from our physical-chemical world to the subatomic world, according to conventional science, looks like this (see fig. 2.2):

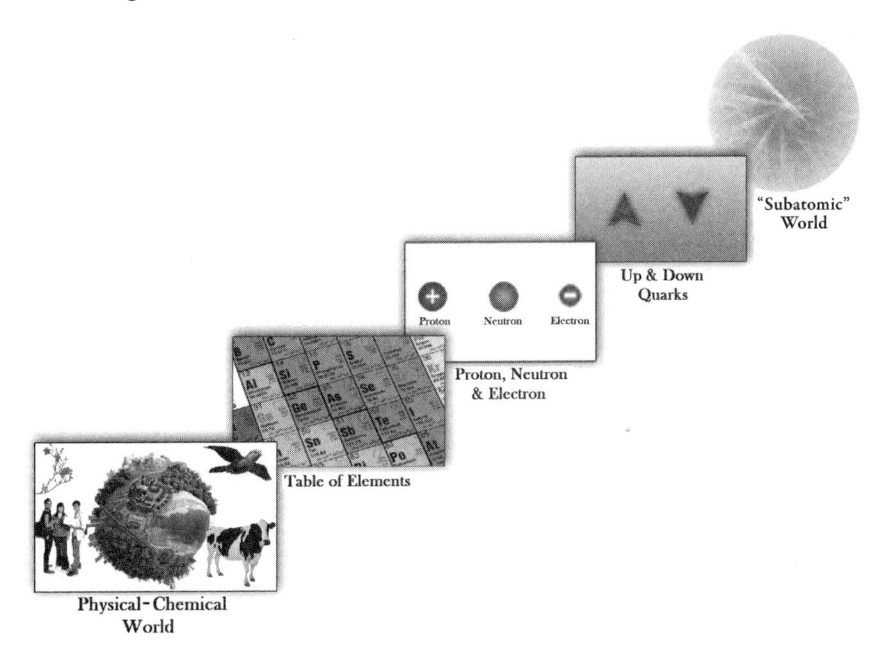

Fig. 2.2. This sequence shows the path, or ladder, from the physical-chemical world to the subatomic world.

Does something *else* exist in the universe to suggest there are more steps on this ladder? And where does subtle energy fit into this picture? Before answering these questions, we need to discuss some anomalies observed by modern science. Most people are impressed by all the

---

*We will see later that unconventional research proved beyond reasonable doubt that subquarks do exist!

technological achievements made possible by the scientific discoveries of the past two hundred years. They have the impression that there are only several small holes in the fabric of our otherwise complete understanding about how the matter of the physical world is constructed and what forces participate in its construction and functioning. However, that is not the case. It's generally recognized that scientists have always encountered and continue to encounter anomalies that cannot be explained using current theoretical and experimental scientific frameworks.

There is an example in our everyday lives that illustrates well how modern science typically responds when discovering anomalous phenomena: Today, all cars with combustion engines have a catalytic converter. This device significantly diminishes the amount of poisonous gas coming from the engine. The action of the catalytic converter is based on the presence of platinum in the device, which causes dramatic changes in the reactions happening in the exhaust gases. *Nevertheless, no change happens to the platinum itself!* It looks like it does not participate directly in the chemical reactions. This phenomenon suggests that platinum contributes some kind of energy to the chemical reaction, because a dramatic change occurs. However, if you ask a chemist what is going on in the converter, you'll get the answer that platinum is a catalyst that stimulates certain reactions resulting in changes in the content of the exhaust gas. But what is the mechanism of a catalytic action of chemically inert platinum? What property of platinum causes the chemical reaction that *doesn't* happen if the platinum is absent? Not only do chemists not have answers to these questions, they don't even ask these questions! They have been taught a definition of catalytic chemical reactions that creates the impression of knowing how these reactions work. But the reality is that existing theories do not offer a precise explanation of the workings of catalysts.

Science ignores the logical idea that *some type of energy* is involved in a catalytic process because no *measurable* energy comes from catalysts like platinum during catalytic reactions. In the same way, modern science continues to ignore many facts that suggest the existence of types of energy—different from electromagnetic energy—which participate in shaping the physical world. However, with the recent discovery of a huge anomaly—namely, dark matter and dark energy—the existence of other types of energy can no longer be ignored.

In 1933, Swiss astronomer Fritz Zwicky was the first to observe that the galaxies in the far reaches of the universe were moving in ways that could not be explained by normal laws of mechanics. Their movement could only be explained with the assumption that the cosmos must contain enormous amounts of an invisible substance that interacts gravitationally but does not react with electromagnetic energy, so could not be measured via light or radio waves.[1,2] Subsequent astronomers estimated that the total mass in a "galaxies plus cloud complex" could be as much as thirty to four hundred times what they could see in their telescopes! They coined this invisible substance "dark matter."

The invention of ultrasensitive radio telescopes gave scientists the ability to view the movement of hydrogen atoms deep in space; not only in the galaxies, but also in the seemingly empty distances between galaxies. Scientists observed that these hydrogen atoms moved as if they were part of a "cosmic gravitational soup."[1] In other words, the void of space is not empty at all. In 1997, observations with the Hubble Space Telescope led to the discovery of the most distant supernova ever viewed. Scientists detected that this dying star "gleams brighter and moves differently than it would if the universe had expanded at a steady rate since the beginning of time."[2]

So, *something* has been affecting the expansion of the universe for billions of years.

In April 2001, three teams of scientists who were calculating the contents of the universe came to the same conclusion: The mass of all of the galaxies, stars, cosmic dust, planets, and everything else "accounts for less than five percent of the universe." The rest takes the form of dark matter (30 percent) and an even more enigmatic dark energy (65 percent) that is causing galaxies to rush apart from one another at an accelerated rate. No matter where we look, dark matter and dark energy keep showing up everywhere.[3] This discovery was not predicted by existing theories of physics. It's no wonder this situation has created a high level of frustration and discussion within the scientific community.

Following is a series of quotes expressing opinions of distinguished scientists that illustrate our current understanding—or rather, lack of understanding—of the dark matter and dark energy phenomena.

> Dark matter is strange stuff. It's all around you, but you can't see it. It's whistling by your ears, but you can't hear it. It is arguably the most important material in the universe, but until recently scientists had no idea that it existed. It will decide the fate of the universe, but we have no idea what it is. How can scientists, after so many centuries, still know so little about the workings of the cosmos? (James Trefil, Professor of Physics at the University of Virginia in *Smithsonian,* 1993)[1]

> Far from shedding light on dark matter, our first experimental glimpses of the elusive stuff have only deepened its mystique. . . . The trouble is that dark matter appears to be different things to different detectors. It appears heavier in one detector than another; it appears more ready to interact in one experiment than another. In the most extreme case, it shows up in one instrument, but not in another—even when both are made of identical material and are sitting virtually next door in the same underground lab. . . . "This is like the story of the elephant," says Feng, referring to the Indian parable in which a group of blind men all touch a different part of an elephant and then compare notes to try to work out what the beast looks like. (David Cline, Professor of Physics & Astronomy at UCLA in *Scientific American,* 2003)[3]

> The discovery of "Dark Energy" is arguably the most important scientific breakthrough of the last 50 years. A full understanding of it eventually could have an effect on our daily lives. But first we have to figure out what exactly Dark Energy is. . . . We call it

> "dark" because we don't directly see it. "Dark" is code for "we have absolutely no clue what it is!" (Meg Urry, Chair of the Department of Physics at Yale University, 2007)[4]

The interesting observation is that mainstream scientists have the same troubles with attempts to understand the nature of dark matter and dark energy as frontier scientists have with understanding the nature of subtle energy. They even use this same parable to describe these troubles.

We can summarize the current understanding of the properties of dark matter and dark energy based on the latest scientific observations and hypotheses:

- Dark matter consists of hitherto undiscovered subatomic or elementary particles. The fact that dark matter is invisible to modern scientific equipment means that these particles *do not interact with the electromagnetic field.*
- Dark energy, like dark matter, is undetectable by today's scientific equipment, which means that it *does not interact, at least in a noticeable way, with the electrically charged particles* (electrons and ions) that are used for detecting the electromagnetic field in scientific equipment. When we say "in a noticeable way," it means that if dark energy is able, for instance, to change the configuration of quarks in the nucleus of an ion, this change would not be registered by any modern equipment.
- It is logical to assume that dark energy interacts with dark matter, and dark matter, in turn, interacts (at least gravitationally) with atomic matter. Otherwise, how does dark energy move galaxies apart? *In other words, dark energy belongs to and operates in the subatomic world.*

Some scientists even suggest that dark matter consists of particles able to make up "the dark equivalent of a hydrogen atom—a dark electron orbiting a dark proton."[5,6,7] If so, then dark atoms can radiate dark light, and vibrations in the dark energy field would be analogous to electromagnetic waves in the electromagnetic field. This assumption raises a "heretical" question about the possible participation of dark matter in the functioning of the human organism. We will discuss this possibility in an upcoming chapter.

The discovery of dark matter and dark energy dramatically changes our view of the universe and our understanding of what modern science is capable of investigating with the equipment it currently has. According to the latest calculations, *only 4 percent* of the universal mass-energy is occupied by atomic matter and electromagnetic energy. This 4 percent is the only part of the universe where scientific equipment is able to take measurements.

We need to realize that the immeasurable substance and energy of the other 96 percent of the universe is right here, on Earth, around us, and as James Trefil said, "whistling by your ears, but you can't hear it."[1] In this situation, we should expect that scientists would be asking questions like:

- Can it be that some unexplained phenomena observed by science are related to the actions or influences of dark energy?
- Does dark energy have something in common with the life force recognized by all Eastern cultures but that, as of yet, remains also not measurable by modern science? In other words, does the chi, or prana, of Eastern philosophy and medicine, or subtle energy in modern terms, present some range of dark energy?
- Being part of nature, are we, human beings, influenced by this all-permeating, universal substance and energy? More than that, is it possible that dark matter and dark energy participate in the construction of the human structure and play a significant role as the life force in our well-being?

Let us remember that, for thousands of years, yogis have spoken about the existence of the subtle bodies of the human structure and their crucial influence on the functioning of the physical body. What are these "bodies" made of? To answer this question, we need to look into the picture of the universe as described in ancient traditions.

In most ancient traditions, our material universe is thought of as having originated from a world of creation that contains all the algorithms for the process of materialization of all things. The process of creation of physical matter, according to such ancient cosmologies, happens through several steps. In each step, a new world is created, with its own laws governing the substances and energies of that world and their interaction. Some modern thinkers are in agreement with this point of view. Dr. Ervin Laszlo writes, "The physical world is a reflection of energy vibrations from more subtle worlds that, in turn, are reflections of still more subtle energy fields. Creation, and all subsequent existence, is a progression downward and outward from the primordial source."[8]

In other words, the universe is constructed of a sequence of worlds, each one made of a successively grosser, or denser, substance. The substance and energy of each of these worlds are created from the previous, subtler world, which many have called the etheric world, or the subatomic world, in terms of modern science.*

According to ancient and modern-day seers, the substances of each world are represented in the human being as sophisticated structures, or bodies, communicating with each other and with the physical body. The substance of the etheric or subatomic world, which is closest to the physical world, makes up the etheric body, which contains the acupuncture meridians and energy centers called chakras.

In light of the current scientific hypothesis about dark atoms, this idea about subtle bodies suddenly does not sound so mystical, esoteric, or off the charts. We know that all periodic table elements, which form the basis of our physical-chemical world, were created inside the stars from the most abundant atom of our world, hydrogen. Is it possible that interactions between dark hydrogen atoms create more complex dark atoms? If so, we must remember: where there are atoms, the formation of molecules is possible; where there are molecules, the formation of sophisticated structures can be expected. In fact, it is logical to assume that structures of the subtle bodies can be made according to the same

*We will talk more about this ancient picture of the universe, and its relation to modern scientific theories, in the next chapter.

general principle, just as our physical body is made from substances of our world.*

The basic components of our physical body are not any different from the components of everything else in the physical world. All inanimate objects and living organisms are made from periodic table elements. What makes living creatures different from each other and from the soil of the Earth is the *difference in their structures.* So, the question becomes, is it not possible that dark atoms create various structures in the world of subatomic, or dark, matter and some of these structures also exist within the human body? Of course, you cannot find a discussion of these heretical questions in modern scientific literature. Nevertheless, we will see from the analysis of some experiments conducted in compliance with rigorous scientific methodologies that these questions have real merit.

Our discovery begins with an analysis of the experiments conducted by the renowned qi gong master Yan Xin, M.D., on changing the half-life, or radioactive decay rate, of the element americium, known as isotope Am-241. Radiation is a natural phenomenon, a fixed quality of certain elements, isotopes, and atoms that we cannot change just as we cannot continue to heat water and at the same time prevent it from boiling. If a human being were to able to change radiation levels, it would be like heating water and preventing it from boiling. It would mean that humans can interfere in natural processes using "dark energy."

Started in 1987, the experiments involving Dr. Xin were part of a long-term research project of the Institute of High Energy Physics of the Chinese Academy of Sciences and Tsinghua University in Beijing. Chinese scientists researched the various effects produced on substances by subtle energy, or chi (qi), projected by Dr. Xin, who is also a chief physician in Chongqing, in the Sichuan province of China. Over time, more than sixty scientific articles were published that report on the many amazing phenomena observed in these experiments.[14] This

*We need to note here that the structure and functions of the subtle bodies and their influence on a human being's physical body have been described in great detail by some seers.[9-13]

research clearly demonstrates the reality of the effect of the subtle energy (chi) on chemical and physical processes in our world.

These experiments were supervised by nuclear physicist Lu Zuyin, Ph.D., a member of the Chinese Academy of Sciences, who wrote a book about this research.[15] Professor Zuyin writes, "The work with Dr. Yan Xin is characterized by careful and rigorous procedures, the use of the most modern and accurate laboratory equipment, and the involvement of other faculty members to form an expert, scientific team."[14] Chinese scientists working in the United States were part of this team. They published an article about the results of the experiments with Dr. Xin, describing in detail all of the equipment and the methods of measurement.[16] Nonetheless, the mainstream scientific community

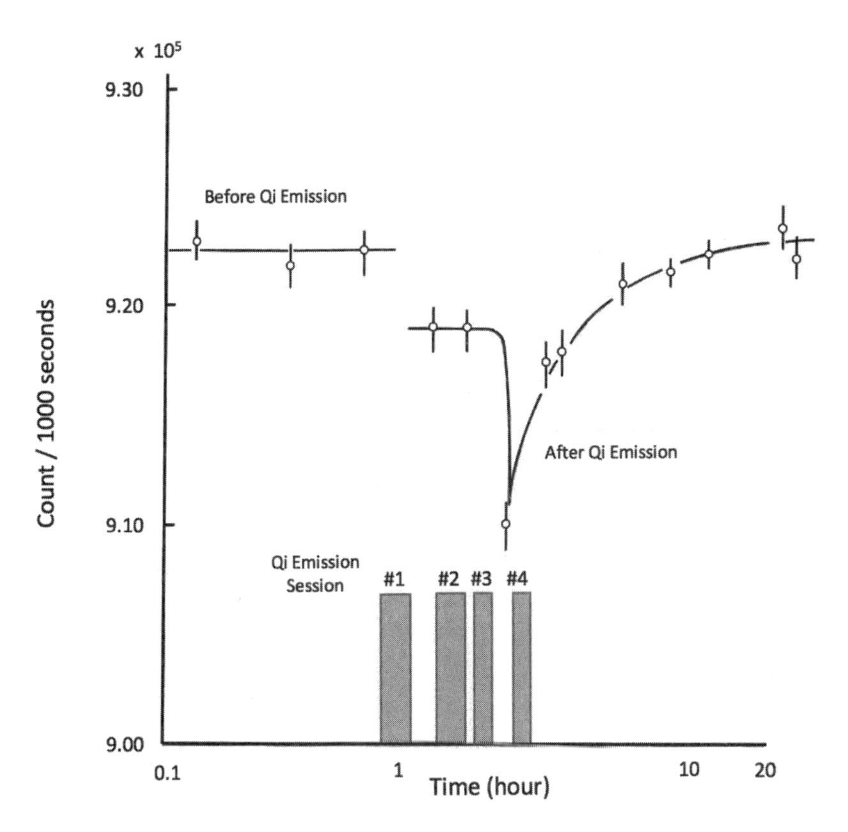

Fig. 2.3. This chart shows the effect of external qi emitted by Dr. Yan Xin during a series of experiments evaluating the radioactive decay rate of Am-241.[15]

totally ignored this outstanding research that, if it were analyzed, could open new horizons for science.

The effect of external qi emitted by qi gong master Dr. Yan Xin about 3 m away on the count rate of gamma (γ) ray decay of Am-241. Note the decreases in count rate during the qi-emission sessions. The error bars in the figure do not include the uncertainty due to the reposition of the Am-241 radioactive source.

The experiments that are important for our discussion—that is, altering the half-life of Am-241—were repeated many times. In these experiments, Dr. Zin projected chi over short, long, and ultra-long distances (from two thousand to ten thousand kilometers) toward the Am-241 radioactive source. The decay rate was measured by counting gamma rays and alpha particles (helium nuclei) emitted by Am-241. Dr. Xin's chi transmission changed the decay rate of gamma rays by up to 10 percent and alpha particles up to 12 percent. Not only was he able to do this from a great distance, but he was able to decrease or increase the decay rate at will! In one series of experiments, the decay rate was changed from +9 to –11.3 percent. In another series, changes were from +6.7 to –12 percent.[16]

Somehow Dr. Xin mentally programmed energy to do the job.

It is important to note that even the strongest electrical or magnetic fields cannot influence the decay rate of radioactive elements. This means that the nature of subtle energy, or chi, is quite different from the nature of the electromagnetic field. Like dark energy, chi cannot be detected by regular scientific equipment. Nevertheless, the chi directed by Dr. Xin dramatically changed something inside the nuclei, which consist of protons and neutrons.

The only logical conclusion is that chi interacts with the particles that make protons and neutrons, quarks, or the even tinier particles, subquarks, which make up quarks. It means that subtle energy, like dark energy, belongs to and acts in the subatomic world. This suggests that subtle energy represents some range of the immeasurable universal energy that modern science calls "dark energy."

According to mainstream science, the decay rate of a radioactive element is governed by one of the fundamental forces in the universe, known as the weak force. The weak force modifies the character of particles; for instance, converting neutrons into protons. During this conversion, the nucleus emits an electron. This process is called beta-decay. If subtle energy can change that decay rate, it means that subtle energy, or chi, is able to alter the weak force, one of the basic forces of the universe.

We don't know whether subtle energy changes decay rates by changing the properties, and therefore the interaction, between quarks and subquarks, or by directly influencing the weak force. In any case, the ability of subtle energy to alter the action of the weak force means that subtle energy may be the *fifth* fundamental force of the universe, in addition to the four fundamental forces known to science, which are electromagnetic, gravitational, strong force, and weak force. In later chapters, we will discuss other experiments confirming this statement.

Another earth-shattering conclusion comes from the fact that Dr. Xin is a famous healer. As both a medical doctor and a qi gong master, he uses his ability to manipulate chi in his healing practice with great success. He studied Western science at both Mianyang and Jiangyou medical schools, and he graduated first in his class at Chengdu Institute of traditional Chinese medicine. When he began his healing practice in 1982, he quickly became known as the miracle doctor, because his treatments were so quick and effective, and he became famous within China for his ability to cure difficult health problems. This led to his collaboration with scientists to document his abilities under controlled conditions.[14]

These healings by Dr. Xin illustrate that the same type of subtle energy—this "fifth force" that affects subatomic particles inside the nucleus—may also affect human health! In the Western world, more and more progressive health care professionals are using various modalities of what is generally called energy medicine or vibrational medicine in their practices. These healing arts that have been known for centuries include laying on of hands, psychic healing, spiritual healing, reiki, and therapeutic touch. There is evidence of laying on of hands as far back as ancient Egypt, as documented in the Ebers Papyrus, dated 1552 BCE. The Greeks used hands-on-healing in their Asclepian healing temples. The Bible references Jesus's laying on of hands and saying, "These things that I do, so can ye do and more," implying that everyone could do this.[17] Currently, some hospitals offer healing touch courses for nurses, and the book *Touch for Health* by John and Matthew Thie[18] is popular among energy medicine practitioners.

In the past two centuries, new healing therapies that use these immeasurable energies have been introduced, such as homeopathy and structured water. Subtle energy may very well be the source of the demonstrated healing effects of such therapies.*

The distant healing effects of prayers, recently validated in scientific double-blind studies,[19] also cannot be explained without consideration of subtle energy. The subtle energy that plays a role in energy medicine must have the same nature as the energy used by Dr. Xin in his experiments. In other words, it is reasonable to argue that energy medicine actually uses the universal energy belonging to the subatomic world that occupies 96 percent of the universe.

There are many experimental results demonstrating the effect of subtle energy on living organisms. In the first decade of this century, Professor Joie Jones (1941–2013) did some astounding research at the University of California, Irvine. Professor Jones investigated "pranic healing" (prana is another term for chi). Jones presented his research at the National Institutes of Health Think Tank Working Group Meeting on BioField Energy Medicine, held in Bethesda, Maryland, in 2006.[20,21] His presentation included the following:

> Here we present a summary of our studies of pranic healing conducted over the past decade. Pranic healers, using techniques established in China thousands of years ago, were able to mediate the effects of gamma radiation on cells in culture. Treatment of cells with pranic healing, both before and after radiation, increased

*We will discuss experiments supporting this supposition in chapter 4.

> the survival rate from an expected 50 percent to over 92 percent. Such enhanced survival rates were independent of the distance between the cells and were not changed even when the cells were shielded from all electromagnetic radiation, including X-rays and gamma rays. In addition, using fMRI, we have measured neuro-physiological changes in subjects treated by a healer. Shielding both the healer and the subject from all forms of radiation had no effect on the results, even when the two were separated by a great distance. Clearly these findings are difficult to explain in terms of the standard scientific paradigm.

This research has an extraordinary message. We know what gamma radiation does to cells: it severely damages them, tearing the DNA to pieces. If subtle energy, or chi or pranic healing, is able to restore cells after this kind of damage, we can understand why traditional Eastern medicine calls this energy the life force!

Like in the experiments of Dr. Xin and Professor Zuyin, the healing effect observed in the research conducted by Jones did not depend on the distance between the healer and the subject. It means that subtle energy does not obey the same law as electromagnetic energy. According to the InverseSquare Law, electromagnetic energy diminishes with distance.

This observation illustrates a fundamental contradiction between the experimentally detected properties of subtle energy and the expectations of modern science about the behavior of any energy field. The strength of all energy fields known to science diminishes as you move away from the source of the field, whether that is an electrically charged object, in the case of an electromagnetic field, or a massive object, in the case of a gravitational field. In the current scientific paradigm, there is no place for the effects of subtle energy to be independent of the distance from its source. Claude Swanson illustrates this void in his book *Life Force, the Scientific Basis,* where he quotes the response of a theoretical physicist to Dr. Xin's experiment on altering the radioactive decay rate of Am-241 from a distance of two thousand kilometers: "There is no way I can believe it, even if I have seen it!"[22]

The amazing, and sad, fact is that, even after discovering that dark matter/dark energy occupies 96 percent of the universe and that modern scientific equipment is only capable of taking measurements on the remaining 4 percent, mainstream science still adheres to the principle proclaimed by British astrophysicist Arthur S. Eddington: "No experiment should be believed until it has been confirmed by theory."

The attitude of mainstream science toward the properties of subtle energy is reflected in mainstream medicine's attitude toward energy medicine. The unexplainable—and therefore unacceptable—properties of subtle energy also show up in observations of its actions in the human body. Frontier researchers have observed how subtle energy functions in the acupuncture meridians. These experiments illustrate how important information on the interaction of subtle energy with the human body can be obtained when scientists follow the principle "replace fear of the unknown with curiosity."

According to traditional Chinese medicine, the acupuncture meridians are the channels through which chi flows, influencing the functions of the physical body's organs and systems. During thousands of years of

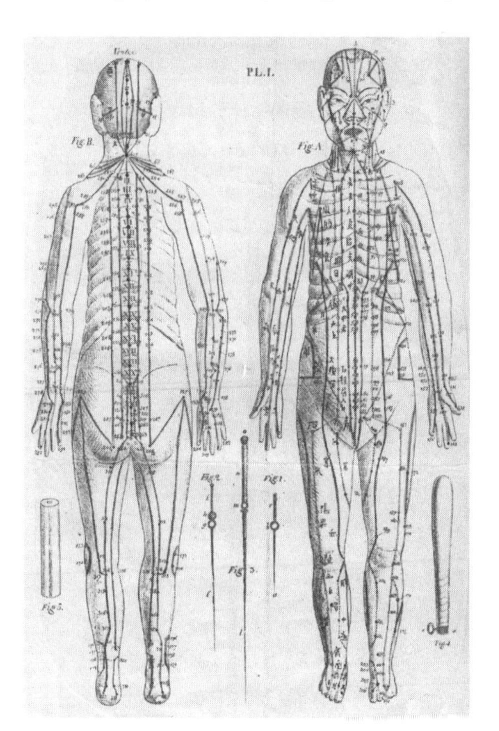

Fig. 2.4. This diagram shows the acupuncture meridians and points in the human body. From *Mémoires sur L'électro-puncture,* 1825, by Jean-Baptiste Sarlandière.

study and practice, acupuncture practitioners developed a sophisticated art and science of using specific points on the meridians for diagnosis and treatment of various illnesses. Acupuncture points are the gateways for moving energy in the corresponding meridians and, consequently, in organs belonging to the sphere of its influence. Acupoints traditionally are activated by needles, heat, or pressure (called acupressure).

Modern acupuncturists also use electrical stimulation of the acupoints. Dr. Reinhold Voll, the founder of electroacupuncture, showed that the electrical conductivity of the body tissues is different in acupoints than in the tissue around them, and it changes depending on the health status of the organ corresponding to that meridian. This fact has led to the development of electroacupuncture equipment that is used both for diagnosis and treatment.

In spite of five thousand years of history and modern confirmation of the efficacy of acupuncture, Western medicine still does not believe in the existence of the acupuncture channel network as a separate system. Again, this is due to the fact that it is not possible to directly measure the flow of chi/subtle energy in the meridians. Research conducted by Professor Jones and his colleagues[23,24] gives us a new and powerful confirmation of the accuracy of ancient perspectives about the acupuncture meridian network.

The researchers used conventional ultrasonic imaging, as well as very short impulses (forty nanoseconds) of high-frequency ultrasound (fifty megahertz), to investigate the anatomy of the acupoints. They found that the acupoints correspond to regions of enhanced elasticity of tissues, which leads to increased ultrasonic attenuation (in other words, the reflection of ultrasound from these regions is diminished). This phenomenon allowed them to receive information about the location and size of the acupoint by sending impulses to the area of the acupoint and analyzing the pulse echo reflecting from the tissue. It even allowed them to re-create a three-dimensional image of the acupoint (see figs. 2.5 and 2.6)!

Their investigations showed that all acupoints are located within the connective tissue. Ultrasonic imaging revealed the following properties of the acupoint:

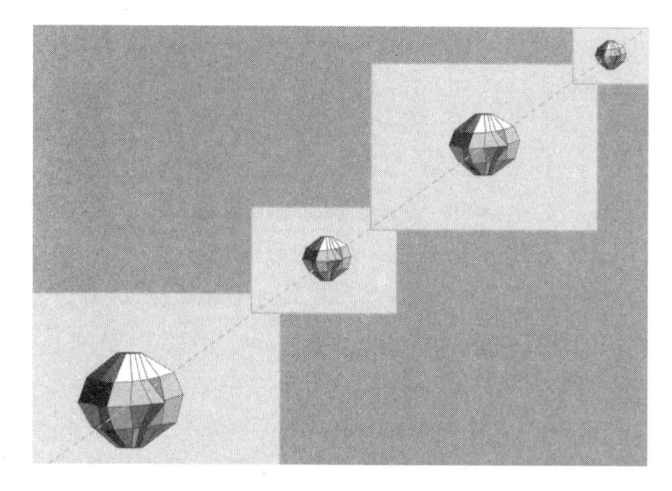

Fig. 2.5. This diagram shows a three-dimensional image of acupoints that were found to be regions of enhanced elasticity.[24]

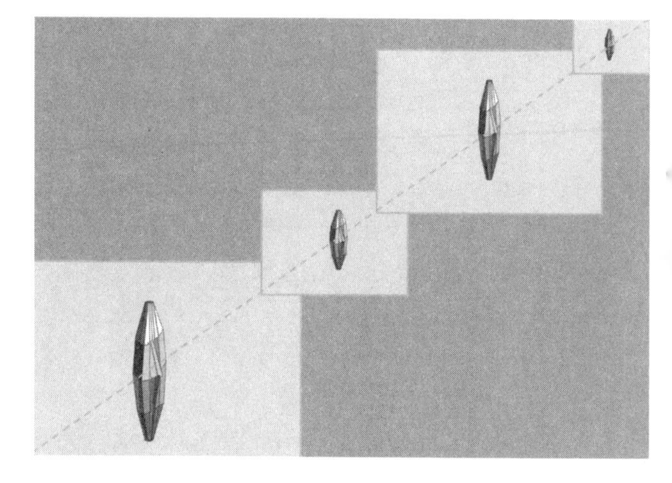

Fig. 2.6. Similar to figure 2.5, this diagram shows another three-dimensional image of acupoints of enhanced elasticity, which were found to be areas of increased ultrasonic attenuation.[24]

- The acupoint changed in size and its center moved slightly over the twelve-day period that the measurements were taken (see fig. 2.7).
- When stimulated by needle or ultrasound, the acupoint twisted and changed shape back and forth, as depicted in the images in figures 2.5 and 2.6.
- The stimulation process was communicated to other acupoints on the same meridian at speeds of five to ten centimeters per second.
- If the meridian was stimulated between acupoints, the stimulation process stopped at the nearest of them.
- An impact made on the surface of the skin or deep into the

muscle tissue—but not at the acupoint—produced no stimulation of the meridian.

Another interesting phenomenon was observed while researchers investigated acupoint BL-67, located on the lateral side of the small toe (see fig. 2.8), which is the last acupoint on the urinary bladder meridian. In traditional acupuncture, this point is considered to be vision related. Using functional Magnetic Resonance Imaging (fMRI), researchers observed excitation of the visual cortex in the brain when point BL-67 was stimulated. This excitation was similar in magnitude and brain location to the excitation produced by stimulation of the visual cortex from flashing light into the eyes. This experiment confirms the ancient acupuncturist's statement that point BL-67 is related to vision.

The biggest surprise came when researchers compared the speed of

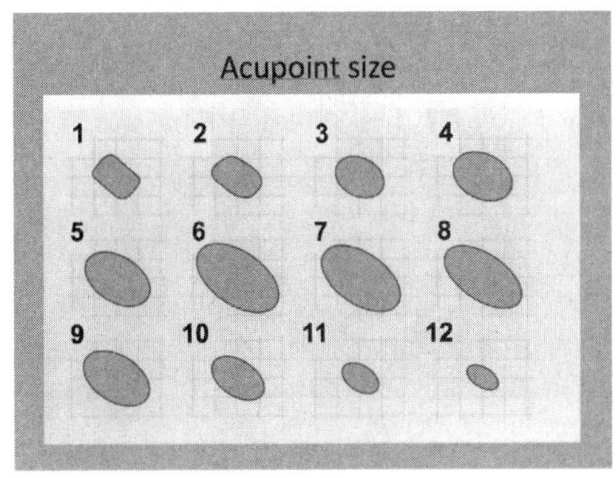

Fig. 2.7. A series of measurements showed a change in acupoint size over a twelve-day period.[24]

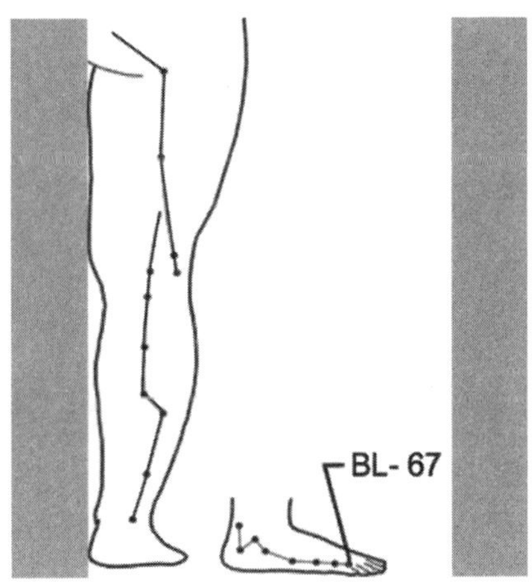

Fig. 2.8. This illustration provides a close-up view of the location of acupoint BL-67.

the signal coming to the brain from the BL-67 acupoint to the speed of the signal coming from the flashing light. The time it took for the brain to respond to flashing light into the eyes was 180 to 200 milliseconds, while the time between the ultrasonic stimulation of acupoint BL-67 and brain response was less than 0.8 millisecond! This means that the signal from the acupuncture point comes to the brain more than two hundred times faster than it does from the flashing light.

We also need to take into account that the signal from acupoint BL-67 travels a much longer distance to the visual cortex of the brain than the signal coming from the eyes. This outstanding research confirms the reality of the existence of acupuncture meridians as a unique energetic network in the body. It also demonstrates the fundamental difference of the processes in the acupuncture meridians and acupoints from any known neurophysiological processes in the body.

It is interesting to note here that researchers reported that "highly sensitive" subjects associated their feelings during these experiments with the flow of energy along the meridians. At the Society for Scientific Exploration Conference, held in Boulder, Colorado, in 2008, Jones made the following conclusion: "The pathway between the acupoint and the brain is different than the nerve pathway. This result is consistent with the concept of the subtle energy field."[24]

This leads to the question: What are the acupuncture meridians made of? If subtle energy, which flows in the meridians, is part of the world of dark matter/subatomic particles and interacts with these particles, it is logical to assume that the meridians are made from this same substance, subatomic particles.* This assumption resonates with ancient science's description of the acupuncture meridians as a part of

*When we speak about immeasurable dark matter as a substance consisting of subatomic particles, it sounds like we expect these particles to be very small; but we actually don't know if they are really smaller/lighter or bigger/heavier than the building blocks of our atoms, protons, and neutrons. The fact that some of the detectable subatomic particles, like the top quark, can have a mass almost equal to that of the nucleus of gold gives us a hint that some of the undetectable dark particles may actually fall outside of what we usually mean by the term *subatomic*.

the etheric body, which is made from pre-physical (or subatomic, in modern terms) substance.

At first glance, these conclusions sound like science fiction. However, let us remember that even mainstream scientists are now inclined to believe that dark matter can consist of complex combinations of particles that form an atomic-like substance. The properties of this substance, and the forces acting there, are largely unknown. Therefore, any new phenomenon demonstrating possible interactions of this atomic-like substance and the energies making up 96 percent of the universe with physical matter, and especially with the human body, should be carefully considered. Changes in the electrical conductivity and elasticity of tissue in the area of the acupuncture meridians indicate that the substance and energy present in the meridians somehow interact with the cells of the tissue.

A follow-up experiment by Jones provides support for the argument that subtle energy propagates in acupuncture meridians. Jones repeated the experiment on registering the signal coming from BL-67 to the visual cortex, with the help of fMRI stimulating the acupoint with a drop of oil that was infused with a subtle energy pattern.* This subtle energy pattern was generated with a kind of technology and developed for stimulation of the urinary bladder meridian. This infusion was done by putting the oil in a specially developed type of generator and subsequently infusing the space with the subtle energy.

The energy infused oil stimulated the acupoint and sent the signal through the meridian, exciting the visual cortex similarly to the excitation produced by needle or ultrasound impulses (see fig. 2.9 on page 50). The only difference was that the effect of the energy infused oil lasted longer than the other types of acupoint stimulations.

In addition to research on the properties of the acupuncture meridians, there have also been investigations into other areas of the human subtle energetic structure, specifically the energy centers called

*In chapter 4, we will discuss features of Vital Force Technology and methods of generating subtle energy patterns developed by the author of this book.

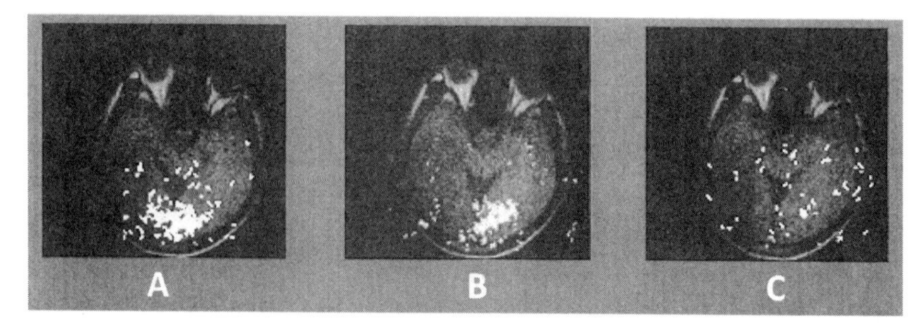

Fig. 2.9. These images of brain scans show the visual cortex region being stimulated during experiments evaluating the effect of energy infused oil on certain acupoints and meridians.

A. Excitation of the visual cortex by flashing light into the eyes
B. Excitation of the visual cortex by oil infused with energy pattern for bladder meridian and applied to acupoint BL67
C. Infused oil applied to random locations on the outside of the foot show no influence

chakras. According to ancient yogic science, the chakras are responsible for energy transfer and conversion between different layers of the energetic structure. The most well-known research was done by Valerie Hunt, Ph.D.,* who found consistent correlations between an aura reader's description of the color changes in chakras and the patterns of electrical signals coming from electrodes attached to the body at the location of the chakras. "After the wave forms were recognizable, there was direct correspondence in every instance throughout all the recordings between these wave forms and the reader's description of primary and secondary colors in the specific chakras."[25]

It is one more indication that changes in the subtle energetic system of the body cause changes in physical characteristics of the tissue. And that means that immeasurable subtle energy has some way of interacting with physical substance.

*More details about Professor Hunt's research can be found in the books of Claude Swanson, Ph.D., and Rosalyn Bruyere.[illegible]

Summary

Experiments conducted in compliance with the rigorous methodologies of modern science allow us to draw the following conclusions:

- Subtle energy acts in and belongs to the world of subatomic particles and therefore might represent a range of what modern science calls "dark energy." Thus, subtle energy may be active in 96 percent of the universe, while known electromagnetic energy acts in only 4 percent of it.
- The nature of subtle energy is very different from that of electromagnetic energy. The strength of its action does not diminish with distance, and it influences the process of radioactive decay, which is not influenced by the electromagnetic force. These facts indicate that subtle energy *might* be the fifth fundamental force of the universe, able to alter the action of the other fundamental forces known to science.*
- Though subtle energy cannot be measured using existing scientific tools, it nevertheless interacts with substances of the physical/atomic world, including the tissues of the human body.
- Undetectable subtle substance of the subatomic world participates in the construction of the human structure as a conduit for the flow of subtle energy and affects the functionality of the physical body.
- One of the most amazing features of subtle energy is that it can be directed, and its action can be programmed, by the human mind.†

---

*To validate this hypothesis, scientists need to open their minds to all the unusual phenomena observed by researchers of subtle energy. So far, there has not been any reaction from the scientific community on experiments, such as Dr. Yan Xin's ability to change the radioactive decay rate by directing subtle energy with his mind.

†This fact opens up new perspectives and possibilities in all fields of science, from physics to biology. It also raises many questions about the necessity of finding adequate methods of exploring this vast area of knowledge, which currently seems extraneous (at best) in Western scientific thinking.

- It looks like it is not possible to describe an energy field that communicates with human consciousness, and which human consciousness can direct, with existing mathematical techniques.
- The phenomenon of subtle energy–human mind communication raises fundamental questions about the nature and origin of human life. For example, is the ability to communicate with this universal energy field an accidental consequence of the evolutionary process? Or is this ability preprogrammed?

We may conclude that the research conducted by modern frontier scientists brings: 1) ever more confirmation of the wisdom passed down in the ancient traditions and practices of Eastern medicine and philosophy, and 2) raises thought-provoking—even paradigm-shaking—questions for both open-minded scientists and truth seekers from all traditions.

In the following chapters we will look at a wide variety of experiments conducted with technologically generated subtle energy that validate many of the statements in the summary given above. At the same time, these experiments demonstrate the tremendous benefits that come from combining ancient wisdom with modern scientific knowledge and technologies for research and practical applications of this possible "fifth force of the universe," subtle energy.

3

Atoms and Strings

Access to Subtle Energy

"Even the glowing grin of the Cheshire Cat implies something else, some larger reality that supports it. Saying that it all gets down to cells and molecules and atoms and quarks, nothing else, seems to leave something out."

—DONALD SKIFF, "THE CHESHIRE CAT" (2007)

EACH DAY, WE WAKE UP in a world where invisible forces constantly permeate our lives. Radio, television, and satellite waves fly around and through our bodies, and we don't question the reality of them—even though we cannot see them! Similarly, when our doctor tells us that we need an X-ray, we don't question this procedure, made possible by radiation that we cannot see, yet which we know works for our benefit.

In the same way, the ancient Eastern traditions of health and knowledge have, for thousands of years, provided people with practical applications of chi, prana, or life force—what we're calling subtle energy today—without them questioning whether such energy exists. People experienced the healing and they readily and consistently felt the effects of the treatments—it was all the proof that they needed!

I am suggesting, here in this book, that there is proof. In this chapter, we will explore *how* subtle energy interacts with the physical world. As discussed in chapter 2, the experiments of Professor Lu Zuyin with Dr. Yan Xin have established a credible foundation for the conclusion that subtle energy interacts with subatomic particles that make up our entire physical universe, and that subtle energy belongs to and acts in the subatomic world.

Considering this, the question arises: Which elementary particles might be responsible for that interaction? We are looking for subatomic particles that are playing a role in the relationships between atoms and the subtle energy field. This would be similar to the role that electrically charged particles play in the interactions of atoms with the electromagnetic field.

We need to remember that all electrically charged particles are not only receptors of the electromagnetic field—which means that they react to changes of the electromagnetic field—but they are also the source of it. For instance, the vibration of an electrically charged particle produces an electromagnetic wave. Analogously, we need to determine the elementary particles that are responsible for delivering subtle energy into our three-dimensional world (in other words, the source of subtle energy). At the same time, we can ask: Are receptors of subtle energy responsible for the effect of it on atomic substances? As we discussed in chapter 2, there is evidence that subtle energy might affect quarks and subquarks inside the nucleus of any atom. Thus, subquarks can be candidates for this role. However, mainstream science does not yet have any evidence of the existence of subquarks. Consequently, we need to carefully investigate what lies beyond the borders of regular science.

Besant and Leadbeater: The Power of Mind

Our investigation begins with some very unusual research that started at the end of the nineteenth century—long before modern science had any idea about the existence of quarks—by two spiritual teachers: Annie Besant (1847–1933), the long-time president of the Theosophical Society, and Charles W. Leadbeater (1854–1934), a former presiding bishop of the Liberal Catholic Church and a fellow theosophist.

In 1907, Besant and Leadbeater observed mass variations of the atoms of neon, argon, krypton, xenon, and platinum. This meant that an element of the periodic table could have atoms with different atomic masses. Only six years later scientists discovered this phenomenon and gave the term *isotopes* to chemically identical atoms with different quantities of neutrons in the nucleus. Besant and Leadbeater also described several elements of the periodic table—*with their correct atomic weights*—that were not known to science at the time (see table 3.1).

TABLE 3.1

Element #	Scientific Name	Published by Besant & Leadbeater	Discovered by Science
43	Technetium	1932	1937
87	Francium	1932	1939
85	Astatine	1932	1940
61	Promethium	1909	1945

Table 3.1. These four elements were first discovered by two spiritual teachers, Besant and Leadbeater, in the late nineteenth century, years before Western science made the same discoveries.

One of the most amazing results of the studies of Besant and Leadbeater was their discovery of the particle—which science later named "proton"—that consists of three smaller particles, two of the same kind and one different. We now know that those small particles are two "up" and one "down" quarks. Scientific confirmation of the existence of protons was completed in 1913. The existence of quarks was theoretically proposed in 1964 and experimentally confirmed in the mid-1990s, almost one hundred years after the research we are discussing here.

Moreover, these visionary researchers described particles we now call quarks as triplets consisting of two constituents, one positive and

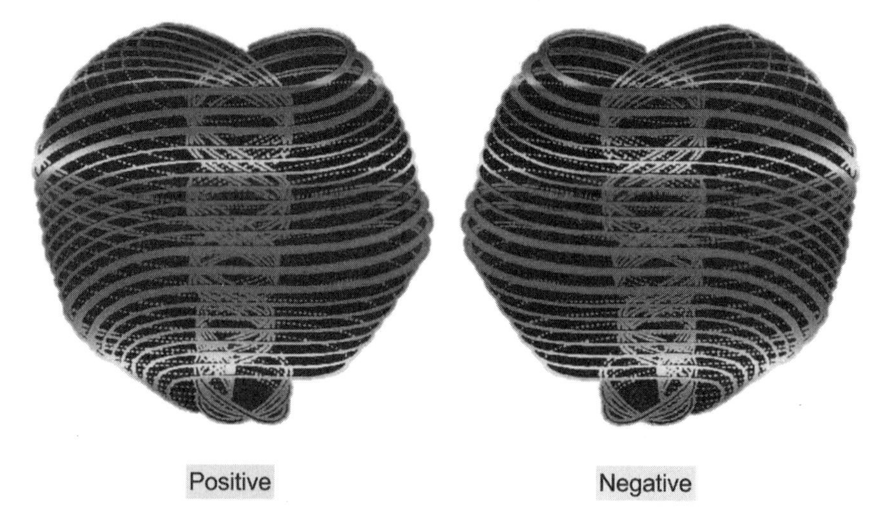

Fig. 3.1. These pictures show UPAs, including positive and negative consituents.[1]

one negative, which they called ultimate physical atoms, or UPAs (see fig. 3.1). In other words, they stated that, as a proton consists of a triplet of two quarks (which modern science later established), quarks themselves are made of triplets of two subquarks. That is *exactly* what current physicists consider as a possibility, although there is still no experimental proof of it.

Besant and Leadbeater applied an ancient "power of mind" technique to investigate the structure of atoms—a technique developed by the second century BCE sage Patanjali, the author of the Yoga Sutras, the most heralded classical text from the yoga school of Indian philosophy.

> Frequent reference is made in the extensive literature of Indian yoga to extraordinary mental abilities ("siddhis") that may be acquired through the practice of yoga. . . . In Aphorism 3.26 of the Sutras,* it is stated that the yogi can acquire "knowledge of the small, the hidden or the distant by directing the light of superphysical faculty."

***Yoga Aphorisms of Patanjali,* translated by S. Prabhavananda and C. Isherwood, in *How to Know God* (London: Allen and Unwin, 1953); also, *The Science of Yoga,* by I. K. Taimni (Adyar, Madras, India: Theosophical Publishing House, 1965).

> The yogi can develop an inner organ of perception that displays "knowledge of the small" in a visual form. . . . The experience of a person in this state . . . is characterized by a vivid, subjective sense of actually being in the microcosmos, of being suspended in space amid particles in great dynamical activity.[1]

Today, we would call what Besant and Leadbeater did "remote viewing" of the microscopic world. Leadbeater had been trained by Indian spiritual masters to develop this special form of magnifying clairvoyance, and he taught this technique to Besant, who later visited Leadbeater's teachers in India to receive additional training.

During the thirty-eight years of their joint research, Besant and Leadbeater observed and described the structure of 111 atoms (among them were 14 isotopes), plus 29 inorganic and 22 organic compounds. The results of their studies were first published in the journal of the Theosophical Society in 1895 and later compiled in three editions of their book, *Occult Chemistry* (1908, 1919, and 1951).

At the early part of the twentieth century—when Besant and Leadbeater were conducting their research—science was still quite far away from discovering the internal structure of the atomic nucleus. We need to remember that even the existence of the neutron was only experimentally confirmed in 1932, just two years before the death of Leadbeater. Thus, even if some open-minded scientists had been present around the time of their research and would have been willing to interpret their results, such an insightful, far-reaching observation would have been virtually impossible to verify. The idea that protons and neutrons are not the ultimate particles of the physical universe had not yet been born, and the existence of smaller particles—constituents of the proton and neutron—would be unimaginable for scientists of that time.

Throughout the nearly four decades of their extraordinary research, Besant and Leadbeater always observed UPAs as fundamental, indivisible constituents of physical matter. Perhaps even more significantly, they saw these rotating and vibrating UPAs as being made of ten non-touching, closed curves, which they called whorls. The whorls in the positive UPA spiral in a clockwise direction, while those in the negative UPA spiral in a counterclockwise direction (see fig. 3.1). Each whorl consists of seven coils (called spirillae). Besant and Leadbeater even depicted how these coils would wind around each other (see fig. 3.2)!

Today, scientists promote the superstring theory of subatomic particles. According to this theory, all elementary particles are different harmonics of the vibration of tiny strings embedded into the homogeneous Higgs field, which is a field of energy that fills the whole universe (therefore, no real vacuum or "absence of anything" ever exists). This vision aligns with the observation of the whorls by Besant and Leadbeater. However, their descriptions of strings, or whorls, are much more sophisticated and precise than the assumptions of modern science.

Of course, one has to be a nuclear physicist to accurately interpret the information and images presented by these two pioneering researchers. This important work was done by a British physicist, Steven M. Phillips, Ph.D., who has made a brilliant and scrupulous scientific analysis of their book, *Occult Chemistry*. Phillips has presented the results in his books, the most recent of them being *ESP of Quarks and Superstrings*.

The analysis accomplished by Phillips shows that UPAs are actually subquarks—the existence of which are currently assumed by

Fig. 3.2. A diagram showing the counterclockwise direction of 5th-, 6th-, and 7th-order spirillae.[4]

modern physics—and the whorls observed by Besant and Leadbeater are constituents of the superstring. Phillips states, "The UPA is the subquark state of the superstring made up of TEN 26-dimensional strings, i.e., the superstring is itself a composite object."

Thus, the superstring is a much more complex construction than existing theories are currently considering. If this is indeed true, it seems unrealistic to expect that the theory of everything will be achieved in the near future, as today's scientists expect. Currently, as the superstring forms used in mathematical models of elementary-particle physics become more sophisticated, the more features of the particles and forces acting between them are revealed. But even with the most sophisticated forms being considered—for example, Vladimir Ginzburg's* toryx model[2]—they are incomparably simpler than the immensely sophisticated subquark state of the superstring, which Phillips determined was observed by Besant and Leadbeater.

These two researchers not only observed and described the structure of subquarks decades ahead of any theorizing about such matters, they also described the forces acting between them as "lines of force" flowing into the heart-shaped depression at the top of a positive or a negative UPA and emanating from their pointed end. According to their observations, these lines of force bind together UPAs in many different configurations (see fig. 3.3 on page 60).

The force that is responsible for binding quarks together, according to modern science, is the so-called colorful force. The existence of this force was theoretically suggested by Oscar W. Greenberg in 1964,[3] in an attempt to explain all the peculiarities of the interactions between quarks. According to this quantum chromodynamics theory, quarks carry, along with an electrical charge, a color charge that can be of three qualities: red, green, or blue.

The fact that Besant and Leadbeater identified what appears to be the string structure of subquark combinations provides evidence that

---

*While Ginzburg's advancements in the theory of micro-world construction are not generally acknowledged by "official" science, we would highly recommend his book for anybody interested in the history and current state of elementary particle science.

they were actually observing what much later became known as colorful force. Certainly, this is a radical notion, but the proof is in what they have clearly described and illustrated.

The illustrations below, taken from their book *Occult Chemistry* and Phillips's book *ESP of Quarks and Superstrings,* provide an example. For a comparison, let us look at what the string model suggests for connections between quarks that create various particles (see fig. 3.4). In the words of Steven Phillips, "the confinement mechanism holding subquarks together is exactly the same as that binding quarks according to the string model."[1] Thus, Besant and Leadbeater confirmed basic ideas of the string model more than fifty years before its creation!

If all of the above, which conveys significant evidence for what the human instrument can achieve, is not enough to boggle your mind, think about this: to observe the internal structure of atoms, Besant and Leadbeater, using the power of their minds, must have had to disintegrate atoms down to the subquark level and then mentally dissect subquarks to see superstrings. This means that they overpowered the strong force, one of the four fundamental forces known to science. In his most recent book, Phillips writes about this phenomenon: "Does this therefore mean that, by the application of psychokinetic forces to

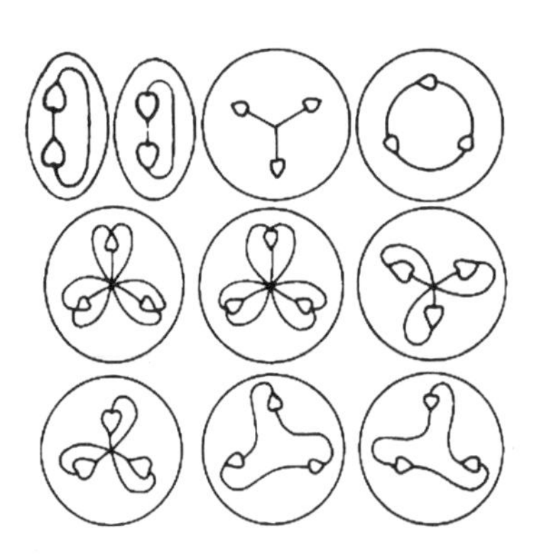

Fig. 3.3. These are examples of hydrogen triplets, or quarks, that were observed in the free, or unconfined state.[1]

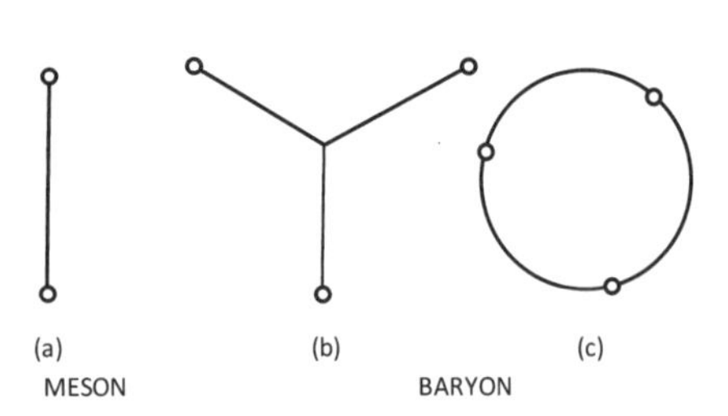

Fig. 3.4. This illustration shows bound states of quarks, according to string theory.[1]

individual subatomic particles, Besant managed what powerful atom-smashing machines may only now be starting to accomplish, namely, reproducing the extremely hot conditions of the early universe by causing the transformation from nuclear matter made up of distinct three-quark bound states (nucleons) to a plasma of free quarks? There seems no alternative explanation of her disintegration diagrams."

Interestingly, "atom smashing machines," like the famous Large Hadron Collider outside of Geneva, Switzerland, cost billions of dollars; in comparison to this, the human instrument is almost free. Can the way Besant and Leadbeater used energy explain how our ancient ancestors moved heavy stones and built structures like the pyramids, which, according to modern science, cannot be done without very heavy equipment that they did not have access to? You can imagine how different modern science would be if only scientists had the courage to seriously consider the results of Besant and Leadbeater's research and Phillips's analysis of their observations.

It is disappointing, yet true to say, that progress in the rigorous scientific investigation of subtle energy could be much faster if scientists would include in their research the possibility of using the human instrument. Such a possibility existed in the beginning of the twentieth century, when some scientists were even members of the Theosophical Society and appreciated the research done by Besant and Leadbeater. Instead, however, the continuous pressure to confirm the existing scientific paradigm has led even leading-edge scientists to conveniently omit the contribution of the human instrument so as to avoid embarrassment, ostracism, termination of funding, or loss of one's position or career.

The case of Francis Aston, a Nobel Prize winner for Chemistry, provides a telling example. Jeff Hughes tells the story in his article "Occultism and the Atom: The Curious Story of Isotopes" published in *Physics World* (September 2003). In 1913, after having discovered what he believed was a new element closely associated with neon, Aston announced his discovery in a paper delivered at the annual meeting of the British Association in Birmingham, England. In his address,

Aston mentioned the relevance of work previously done by theosophists Charles Leadbeater and Annie Besant. He stated the pair had used "theosophic methods entirely unintelligible to the mere student of physics" to determine the atomic weight of all the known elements and some unknown (at the time). They ascribed an atomic weight of 22.33 to one element and called it Meta Neon. In fact, this is the name Aston chose for the new element with an atomic weight of 22, which he had scientifically discovered and which was closely related to neon (atomic weight = 20.2).

Aston went on to receive the Nobel Prize in 1922 for his work with isotopes, but in his Nobel lecture and in a textbook he wrote that year, Aston, as Jeff Hughes writes, "reconstructed the history of his own work to make the link between neon-22 and isotopes seem straightforward . . . dismissing the language of 'meta-elements' . . . as a false path [and] . . . all reference to occult chemistry was eliminated." Aston's rewriting of scientific history disconnected his theory "from a set of ideas that he and his new mentors would have found embarrassing."

In just the same way, who knows how many significant discoveries have been marginalized by those scientists lacking commitment to the truth laid plainly before them? How many important findings are even now being suppressed because of the lack of truly open-minded scientific inquiry displayed by scientists like Francis Aston? One can only guess.

Unfortunately, today's scientists are scared of the word *ESP,* extrasensory perception, or they have not given credence to those studies which *have* provided scientific evidence for the wide-reaching power of the mind. Instead, most have gotten used to believing only what instruments perceive, and they simply don't believe in the potential of human perception.

A proper and thorough discussion of all aspects of the observations made by Besant and Leadbeater, deep inside the subatomic world, as well as Phillips's comprehensive scientific analysis of their results, is beyond the scope of this book. Here, we will concentrate on the main features and properties of the UPAs, the ultimate physical atoms that were established as a result of their observations. A basic understanding of this will be important for achieving our goal to suggest a feasible hypothesis for a mechanism of interaction between subtle energy and physical matter.

Main Features of UPAs and the Forces Acting Through Them

The two UPAs, positive and negative, have been described by Besant and Leadbeater as "alike in everything save the direction of their whorls and of the force which pours through them. In the one case, force pours in from the outside from fourth-dimensional space, the astral plane, and passing through the [UPA],* pours into the physical world. In the second, it pours in from the physical world, and out through the [UPA], into the outside again, i.e., vanishes from the physical world. The one is like a spring from which water bubbles out; the other is like a hole, into which water disappears."[4]

Based on his analysis of the UPAs and their interactions, Phillips writes, "In the modern terminology of fluid mechanics and particle physics, positive and negative UPAs are, respectively, 'sources' and 'sinks' of hypercolor gauge forces, i.e., respectively, positive and negative magnetic monopoles. . . ." Subsequently, he continues, "[T]he UPAs making up a proton or neutron are, as subquarks, bound by strings both inside a constituent quark and externally to subquarks in the two other quarks. This means that they are monopole sources and sinks of two different types of hypercolor gauge fields, one responsible for confinement of subquarks in quarks, the other responsible for confinement of quarks in baryons, like protons and neutrons."

In other words, UPAs are the magnetic monopoles predicted by superstring theory that current science is still looking for. According to Phillips's analysis, some components of the force pouring through

*In *Occult Chemistry,* third edition (p. 13), the authors referred to UPAs as "Anu." For the sake of consistency, in this book we have replaced the name "Anu" with the name "UPA," including in all of the quotes taken from the earlier book.

UPAs (and connecting them, and thus all physical matter, with the higher and multiple dimensional, which Ervin Laszlo calls "multiverse") create the strong force responsible for the existence of all periodic table elements. Consequently, they are responsible for the existence of all chemicals that make up our world; that is, the physical basis of the multiverse.

Furthermore, with a thorough examination of Phillips's analyses of Besant and Leadbeater's research, we see that subquarks/UPAs are responsible for the creation—in our three-dimensional, atomic world—of three of the four fundamental forces of nature acting in this world: the strong, weak, and electromagnetic forces.* (For those who are interested to know more about the logic behind this conclusion, please refer to appendix A.)

Now, what about the energy field that the ancients throughout the world have called life force and that we call subtle energy? Is there some connection between UPAs and this force? In the Leadbeater-Besant book, *Occult Chemistry,* they write, "The [UPA] can scarcely be said to be a 'thing,' though it is the material out of which all things physical are composed. It is formed by the flow of the life force and vanishes with its ebb. The life force is . . . the force of which all the physical plane forces are differentiations. When this force arises in 'space'—the apparent void which must be filled with substance of some kind, of inconceivable tenuity—[UPA] appear: if this be artificially stopped for a single [UPA], the [UPA] disappears: there is nothing left."[4]

As we can see from Phillips's analysis, differentiation of the energy that Besant and Leadbeater define as life force into the physical plane forces happens by means of the interaction of that force with the UPAs. This assumption about life force/subtle energy entering all physical substances, and therefore the surrounding space, by means of vibrating whorls of ultimate physical atoms leads to a whole chain of conclusions

*The origin of the gravitational force remains a puzzle. According to ancient knowledge, it acts in more than the three-dimensional universe. In other words, the gravitational force influences substances of more subtle constitution existing in the multiverse (e.g., dark matter).

about alleged properties of subtle energy and the laws of its interactions with other forces and with physical matter.

Though validity of this basic assumption cannot be confirmed directly, we will discuss in chapters 5 through 7 some fundamental experimental facts about subtle energy that testify in favor of this assumption.

Subtle Energy Signatures (Patterns) of Substances and Their Properties

When we look at the nucleus of an atom as a combination of the UPAs, each individual atom from this perspective looks like a big molecule made of many UPAs. For instance, the nucleus of oxygen consists of 16 protons and neutrons; each of these is made from 3 quarks, and each quark has 3 UPAs. Thus, oxygen's nucleus contains 144 UPAs. The nucleus of gold, which has 197 protons and neutrons, contains 1,773 UPAs.

If, as discussed above, each UPA is a source (positive UPA) or sink (negative UPA) of subtle energy, then each atom presents a unique distribution of the subtle energy flow, depending on the amount of positive and negative UPAs and their geometric configuration in the nucleus of a particular atom. In other words, each atom of the periodic table has its own unique subtle energy pattern. When atoms combine into structures such as molecules or crystal lattices and create various substances, subtle energy patterns of those substances are even more sophisticated and unique.

Accepting our hypothetical model, which states that the UPAs are conduits of subtle energy from the multiverse to the three-dimensional physical reality, leads us to identify the following general properties of the subtle energy patterns of substances:

- Atoms within a substance work more as conduits of subtle energy than they do as a source or capacitors of it. As a result, subtle energy behavior can appear very different from the behavior of the electrical field of the same substance. For instance, we cannot observe any discharge of a subtle energy pattern of a substance, as the flow of energy will continue until the substance is changed. Later, in chapters 5 through 7, we will present many experimental confirmations of this phenomenon.

- External subtle energy can, under certain conditions, influence the configuration of UPAs inside the nuclei of atoms of a substance, which changes the subtle energy flow distribution, naturally affecting the properties of the subtle energy pattern of this substance.

It is important to notice that, since the total electrical charge of the nucleus stays unchanged, the alteration of a substance's subtle energy pattern can happen without any noticeable changes in the chemical properties of the substance. In other words, chemically identical substances can have noticeably different subtle energy patterns.

In chapters 4 through 7, we will present experimental confirmation that specific subtle energy patterns technologically infused into substances can alter energetic properties of these substances.

It is possible that specific external subtle energy patterns may change the behavior of whorls in the UPAs, which might result in the change of the subquarks' interactions. This can lead to the redistribution of electrical and colorful charges inside the nucleus; the former effect can alter electronic orbits in the atom, while the latter effect can influence the weak force. This makes it possible to explain why subtle energy is able to change the course of chemical reactions, as well as influence the electronic and vibrational spectrum of substances.[5,6,7] It also helps explain the altering of an isotope's radioactive decay rate in the experiments conducted with Dr. Xin, discussed in chapter 2.

It is important to recognize that Besant and Leadbeater were really the pioneers of superstring theory. Their groundbreaking work preceded by many decades the work of today's nuclear physicists, who test the predictions of particle physics theories with multibillion-dollar atom smashers.

In conclusion, there are two sets of profound results that were achieved: first, Besant and Leadbeater in their magnifying clairvoyance observations of the subatomic structure of substances; and, second, the rigorous scientific analyses and interpretations of these results by Steven Phillips, Ph.D. Together, in this author's opinion, these open unimaginable horizons for science in the understanding of the workings of the universe and our human role in it.

All of the achievements discussed here also point to unprecedented possibilities for scientific research on the interpretation of phenomena previously considered mystical or superstitious. The results of this author's research using technologically harnessed subtle energy, which had been inspired and directed by the material discussed above, will be presented in chapters 4 through 7.

We will see how unusual phenomena—at least from the viewpoint of mainstream science—that have been observed in a wide variety of experiments (including involvement of the human mind in the process of programming subtle energy effects) might be illuminated, explained, *and practically used,* with the help of the hypotheses presented above, regarding the interaction of subtle energy with physical matter.

PART 2

Experiments and Application of Subtle Energy

4

Vital Force Technology

BY UNDERSTANDING how subtle energy interacts with substances of the physical world, we can discover how to use modern technology to harness it. This opens a new realm of seemingly limitless possibilities. We will begin to see how subtle energy can be used to improve and enhance many areas and dimensions of our lives: health, agriculture, and even the environment.

We have seen that mainstream science speculates about the existence of subquarks as the fundamental particles that make up all substance of the physical world. This emerging perspective matches Besant and Leadbeater's observations of UPAs that we discussed in the previous chapter. According to these researchers, two UPAs are the entrance and exit doors connecting our world with the life force—the energy of the pre-physical/subatomic world that we call "subtle energy." As we mentioned in chapter 3, the two UPAs, positive and negative, have been described by Besant and Leadbeater as "alike in everything save the direction of their whorls and of the force which pours through them. In the one case, force pours in from the outside from fourth-dimensional space, the astral plane, and passing through the [UPA], pours into the physical world. In the second, it pours in from the physical world, and out through the [UPA], into the outside again, i.e., vanishes from the physical world. The one is like a spring from which water bubbles out; the other is like a hole, into which water disappears."[2]

As follows from this description, the energy patterns of substances, transmitted by their UPAs, can be influenced by electromagnetic vibrations. This happens because of the electrical charge residing in quarks, which are composed of the UPAs. Electrons, according to Phillips's analysis, being just different states of the UPA (for details, see appendix A), also carry an electrical charge, as well as whorls (chapter 3).[3] The electrons interact with the electromagnetic field, while the whorls interact with subtle energy. As a result, vibrations of an electrically charged particle stimulated by an external electromagnetic wave can change the interaction of whorls with subtle energy, resulting in some changes of the energy pattern of the substance. This opens a way for us to create various subtle energy patterns.

Although the suggested picture of the interaction between electromagnetic energy and subtle energy is a hypothetical one, the experimental facts we will discuss in following chapters unquestionably prove the existence of this interaction.

More than twenty-five years ago, I began to experiment with the generation of subtle energy patterns using an electron-ion plasma, which was created by an electrical discharge in various gases. Vibration of the current flowing through the plasma and control of the parameters of the vibration were performed by specifically designed electronic equipment.

Taking into account that plasma is a nonlinear medium, current versus voltage modulation is necessary to escape an appearance of the harmonics of the modulation frequencies. During the years of research, this equipment was continuously modified in accordance with the progress in electronics and computer sciences.

A general schematic of the VFT we are now using is presented in figures 4.1 and 4.2 on page 72.

Vibration in plasma generates electromagnetic waves. According to our model of the interaction of the electromagnetic field with subtle energy, vibrating plasma should also generate the flow of subtle energy. If we change the gas-forming plasma, while keeping all parameters of its modulation the same, electromagnetic radiation of the generator will

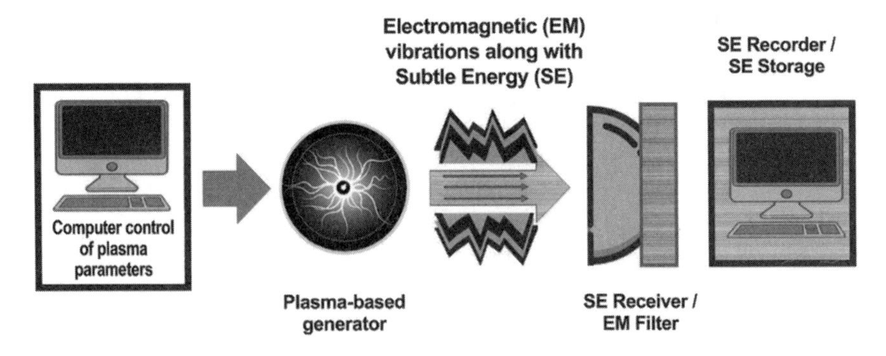

Fig. 4.1. This diagram illustrates the process of generating and recording subtle energy.

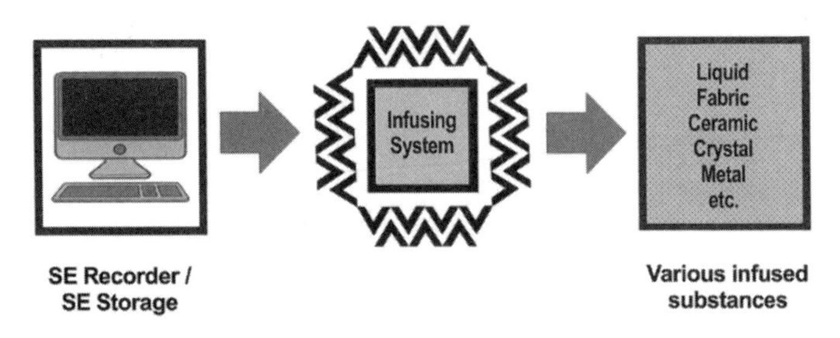

Fig. 4.2. This second diagram shows the steps involved in VFT's storage and infusion system.

remain unchanged. However, this is not true for subtle energy. As we established in the previous chapter, subtle energy patterns of various substances can be significantly different from each other. That's why precisely identical vibrations of plasma made of different gases, for instance, inert gases—helium, neon, argon and krypton—create subtle energy patterns with *different* properties.

This was confirmed during the original experiments with VFT by observations of these patterns' effects on both acupuncture meridians and the autonomic nervous system as measured by electroacupuncture according to Voll (EAV) devices and heart rate variability (HRV) tests. This outcome supports the conclusion that each periodic table element is characterized by a unique subtle energy pattern.

As electromagnetic waves keep interacting with—and changing—subtle energy patterns, it is important to separate the electromagnetic radiation while recording the subtle energy patterns. Special types of receivers were developed that are able to accept a subtle energy flow while filtering electromagnetic radiation up to about –70 dB (in other words, a reduction of background noise by 10 million times) at the input of the recording system. Additional filtering installed into a recording device added another –70 dB, bringing total suppression of the electromagnetic noise's power to –140 dB!

Along with the requirement of electromagnetic radiation noise suppression, we found that when using a digital type of the subtle energy's recording, high levels of a sample rate (192 kHz and higher) and resolution (32 bit and higher) need to be used to escape distortion of the energy patterns being recorded. Doing this using a regular CD recorder's parameter, and especially when using compressed formats like MP3 and others, exhibited an alteration of the original energy patterns to a level that couldn't be effectively used for the intended applications.

Finally, we need to emphasize the relevance and influence of energetic conditions—specifically those of the environment where recording is carried out—on the quality of the recorded subtle energy patterns. This includes the energetic state of the people producing the recording.

The effects of energetic pollution on the outcome of various scientific experiments will be discussed in chapter 7 and appendix E. The most obvious example of an experimental outcome that might be affected by environmental noise is the process of recording subtle energy by electronic equipment. Here is a good analogy: the quality of music you hear on your car's AM-FM radio while driving is affected by the electromagnetic noise taken in by the car's radio receiver (the static you hear or even the dead silence when there's too much interference).

That is why a VFT recording system is located in a room protected against the electromagnetic pollution that diminishes any subtle energy pollution related to it. Also, to counteract any hard-to-detect energetic pollution, this room is periodically cleaned using a special energy pattern created for this purpose (for more on this, see chapter 7 and appendix E). Additionally, people performing the recording are experienced meditators capable of determining and correcting their own energetic state.

Methods of Creating Specifically Targeted Subtle Energy Patterns

1. Using a VFT Generator as a Source of Subtle Energy

During the years of experimentation with the VFT generator, many important, practical features of the plasma-based method of subtle energy pattern formulation were discovered. First of all, we noticed that it is possible to determine ranges of modulation frequencies that produce energies affecting certain levels of the human body system. For instance, a frequency range from about 150 Hz to 300 Hz influenced mostly physiological processes, 300 Hz to 450 Hz affected mental activity (including brain wave intensity), and 450 Hz to 600 Hz influenced the emotional state. Going up in the modulation frequency, it was possible to find higher ranges that began to repeat the physiological-mental-emotional sequence of influences.

Of course, this data is only a first small step into the vast field of subtle energy's properties and puzzles. The frequency ranges pointed out here, for example, significantly depend on the colors of gases used to create plasma and the filtration of the light emitted by it.

Along with the regular scientific equipment used during our experiments—HRV, brain mapping (EEG), EAV, and so on—we also used a "human instrument," in the form of extremely sensitive people, so-called medical intuitives. We developed special protocols of testing to maximize the accuracy of that unusual kind of testing.

Furthermore, we discovered that by sweeping the frequency of modulation, it is possible to create energy patterns that are biologically highly active. For instance, one of the patterns, called stress relief, was created by sweeping the frequency of the plasma vibration in the range of 250 to 400 Hz, using a specific combination of inert gases. The results of testing this energy pattern strongly corroborated that it influences the sympathetic/parasympathetic nervous system balance particularly well, which encouraged us to conduct an experiment on mice to objectively test its properties (see chapter 6 and appendix D for more on this).

One of the peculiar features of these sweeps is that their effects can be dramatically changed, if the time or width of the sweeping is altered by just one second or one hertz. In other words, the parameters of a sweep need to be determined with high precision.

Another example of the unique possibilities that the VFT generator offers is the ability to create and amplify a subtle energy pattern using modulation frequencies equal to the frequency of the Schumann resonance and its harmonics. This name was given to the electromagnetic standing waves created between the surface of the Earth and the ionosphere of our planet. They are generated by lightning discharge, which is a phenomenon that is happening somewhere in the Earth's atmosphere about fifty times every second (we will discuss this more in chapter 7). This phenomenon may very well have a global effect on all life on Earth. Thus, the creation and recording of the Schumann resonance's subtle energy wave opens the door for rigorous scientific research of its effects on living organisms.

Phantom Atoms

I have been greatly intrigued by the challenge of creating energetic counterparts of the cornerstones of our physical world: atoms of the periodic table of elements. As I thought of this possibility, I remembered a principle of ancient cosmology: lower worlds are created by multiple divisions (by two) of frequencies of all vibrations in upper worlds (for an example, see *A Kabbalistic Universe* by Z'ev ben Shimon Halevi[1]).

The main vibrational characteristic of atoms is the frequencies of their electrons' transition from one energetic level (called an orbit) to another. Those frequencies are located within the range of light frequencies: infrared, visible, or ultraviolet. An important characteristic of electronic transitions is the intensity characterizing how strongly they interact with light waves.

Combining the above-mentioned principle of ancient cosmology with modern experimental data on frequencies and intensities of electronic transitions in atoms, energetic blueprints related to many atoms were created with the help of the VFT generator. For this purpose, frequencies of several of the most intense electronic transitions in an atom were used. We chose their sub-harmonics belonging to the range of audible sound frequencies to modulate plasma. We found that energetic phantom atoms created using these principles, when tested by professional health care practitioners, demonstrated effects on the human body similar to the effects known to arise from their chemical counterparts. The phantom atom of lithium was found to work against depression, the energy pattern related to germanium was reported to help the immune system, and the pattern related to oxygen was discovered to help assimilate significantly more oxygen for people walking at high altitudes. The latter was tested with a standard oximeter.

All of these discoveries pave the way for previously unimagined possibilities for using phantom atom subtle energy patterns in the field of energy medicine. It means that we can replace (or diminish the amount of) chemical elements used for treatment, and thus avoid their side effects. A good example of this is lithium, which works well as a strong anti-depressant but is also known to have a strong negative effect on the liver.

The Five Elements Principle

Our experiments also led to the unexpected discovery that it is possible to generate subtle energy patterns with properties described by traditional Chinese medicine as the elements fire, earth, metal, water, and wood. This was accomplished by determining the appropriate frequency ranges of plasma vibrations, proper parameters of the plasma, and filtration of the light coming from the VFT generator. As the base for finding frequencies dividing ranges with different elemental properties, we took the statement from traditional Chinese medicine that each of the elements is associated with two numbers. For water, those numbers are 1 and 6; for wood, they were 3 and 8; for fire, 2 and 7; for earth, 5 and 10; and for metal, 4 and 9. From these numbers, it is possible to create an analog of a Fibonacci sequence, where each number in the sequence is the sum of the two numbers that precede it. For water, this sequence starts with 1, 6, 7, 13, and so on; for wood, it starts as 3, 8, 11, 19, and onward; for fire, 2, 7, 9, 16; for earth, 5, 10, 15, 25; and for metal, 4, 9, 13, 22, and so on.

Using these numbers as frequencies, we created a particular pattern for each of the elements. Frequencies of all five elements in audible range are presented in appendix B. One can see that in each series of these frequencies, corresponding to the same level, numbers are consistently growing in the sequence, as in the earth-fire-wood-water-metal sequence, or diminishing in the opposite direction, each time corresponding to the supporting sequence of the elements without overlapping.

It is well known in traditional Chinese medicine that each of the acupuncture meridians possesses a specific elemental quality. Therefore, using electroacupuncture measuring equipment, based on Dr. Reinhard Voll's EAV devices and methodology, it is possible to determine major elemental qualities of subtle energy patterns by testing their effects on various acupuncture meridians. We infused subtle energy patterns into oil and then put a drop of the oil on an acupuncture meridian. Our experiments with several rows of frequencies, presented in appendix B, show that subtle energy patterns of exact frequencies corresponding to the particular element (such as 280 Hz for fire and 364 Hz for water) produced a stronger effect on the meridians with corresponding elemental properties than patterns of other frequencies in the same row. We need to add that patterns generated with exact elemental frequencies acquire more pronounced elemental properties when light coming from plasma is filtered with the color corresponding to the particular element; in other words, white for metal, blue for water, green for wood, red for fire, and yellow for earth.

With this knowledge, it is possible to create energy formulas for specific applications using methods suggested by traditional Chinese medicine, including the recommendation of combining herbs. As an example, we can refer to the well-known traditional Chinese medicine algorithm caldron that is intended for improving digestion. It prescribes creating a slow fire, combining wood, water, and fire elements in specific proportions. Energy patterns with the frequencies in the range below 300 Hz (see appendix B) of these three elements were created and infused into a solution of water and trace minerals (TMS) with the help of our VFT system (see figures 4.1 and 4.2). Then, TMSs with different elemental energy patterns were mixed in the optimal proportion and sent to health care practitioners for testing. All of them found that the energy formula called GI Aid very positively influenced people with various digestive problems. Later we added the phantom atom energy patterns of chromium and vanadium, which are known for helping blood sugar regulation, to this formula. Since then, GI Aid is reportedly being used successfully by health care practitioners in several countries. This example demonstrates the effectiveness of combining subtle energy patterns to create sophisticated formulas targeted for specific applications.

2. Mapping Subtle Energy Patterns of Substances

As we discussed in previous chapters, each physical object possesses a unique subtle energy pattern. This also relates to substances used for health maintenance, such as herbs, vitamins, and amino acids. Energy patterns of such substances can be copied and recorded using VFT equipment (see fig. 4.1). For this purpose, substances are simply put into the receiver and the recording system is switched on for several minutes.

Today, there are many computerized energy medicine devices that use uploaded energetic imprints of various remedies for healing.

Usually, the recording of these imprints (we call them subtle energy patterns) is accomplished using an approach similar to the recording of regular sound. While this method for recording subtle energy patterns is possible, we have found in our research that many of the energy patterns of the remedies recorded into computerized devices are significantly (and sometimes dramatically) less effective than the energy patterns of the original remedies. We think this loss happens due to the strong electromagnetic noise associated with the recording devices.

We have investigated this phenomenon for several years in association with William Lee Cowden, M.D. Dr. Cowden is a well-known teacher of integrative medicine who uses a unique kinesiology test to evaluate the effectiveness of subtle energy–based remedies. In our joint research, we have conducted double-blind experiments to compare the quality of recorded energetic patterns that were infused into a TMS using VFT equipment.

The research with Dr. Cowden helped us to develop a very low-noise and high-resolution system for copying the energy patterns of substances. We developed this system until Dr. Cowden didn't see a difference between the quality of the remedy's recorded energy pattern and the energetic quality of the remedy itself. After developing an effective method of no-distortion copying for subtle energy patterns of substances, we began to work on finding ways of enhancing and modifying the physiological effects of these patterns. One of the effective methods to achieve the latter was found while copying the energy of a substance infused with the energy pattern of another substance.

Surprisingly, there are times when the resulting combined energy pattern might be quite different from the patterns of both of the components. We created the Zen pattern through copying the energy pattern of green calcite, infused with the Clean Sweep subtle energy formula, made for cleaning environmental energetic pollution (for more on this, see chapter 7 and appendix E). Subsequently, we discovered that when infused into a water mineral solution, Zen could be effectively used as a support for more swiftly achieving a deep meditative state. This result has been confirmed by many experienced meditators. However, neither the Clean Sweep nor the energy pattern of green calcite possesses this quality.

We have also found that it is possible to enhance or modify the properties of energy patterns of substances during their copying through using the power of intent of the people performing the recording. We have named these methods of recording "mapping," to distinguish them from the direct copying of the energy patterns of substances.

3. Healers' Intent

The research by Dr. Xin and by Besant and Leadbeater, as described in the previous chapters, leaves little doubt about the human mind's ability to affect physical matter and even the fundamental forces acting in the material world. It is important to note that these abilities can be learned and enhanced by special mental exercises, some of which were developed millennia ago. This has been confirmed by the experiments of modern frontier scientists. For example, Dean Radin, Ph.D., writes in his book *Supernormal* that, in one of his many experiments with changing the interference pattern of light going through two slits, experienced meditators "obtained combined odds against chance of 107,000 to 1. Non-meditators obtained results close to chance."[11] Radin also discusses what is known about ancient yogis' techniques of achieving higher states of mind that lead to gaining supernormal powers, called *siddhis,* which is Sanskrit for "accomplishment" or "perfection."

We also need to add the experiments on using prayers for healing. The effectiveness of prayers for healing has been demonstrated not only on humans but also on animals.[14] As the placebo effect is not applicable to animals, we are left to conclude that the results of these experiments are totally objective.

These phenomena attract more and more attention from frontier scientists and writers, and many books have been written on the topic of the mind-matter interaction.[4–13] In these books, you can find a great deal of experimental evidence demonstrating the mind's effects on various processes observed by modern science, including influencing the outcome of random number generators[4] and altering the interference pattern of light coming through two slits.[6,9,11] With all this evidence of the ability of the human mind to alter processes in inanimate objects, as well as in living organisms, a very important question arises: Is it possible to scientifically determine the mechanism of the mind-matter interaction?

We could say, "Well, it is pretty obvious that all observed effects of the mind on physical matter and living creatures are possible due to the force we call here 'subtle energy.' After all, in all of the experiments involving Dr. Xin, Chinese scientists use the term 'external chi,' meaning that Dr. Xin somehow directs chi energy existing outside of his body, and 'programs' this energy with his mind to produce the intended effect."

Our common sense tells us this interpretation is valid. However, from a rigorous scientific point of view, there is no proof that chi/subtle energy is the carrier of intent of Dr. Xin. The reason is that we do not have an instrument suitable for measuring a flow of subtle energy or any changes in the subtle energy field indicating that Dr. Xin manipulates this energy. In all such experiments, scientists can only observe the source of the influence—in this case, Dr. Xin himself—and the result of the influence. However, what is happening in the space between the source and the recipient of the effect is totally unknown, at least from the point of view of Western science. This is exactly the reason that the wide range of so-called psychic phenomena—including distance healing—is considered by many to be miraculous.

Can we find out whether something is happening in the space between the source of the effect and the target? In other words, can we demonstrate there *is* a carrier of the effect that is delivering this effect to the target? Below, we present evidence that, by using VFT, it is possible to acquire experimental proof that miracles of mind-matter phenomena are happening via the interaction of subtle energy with the human mind.

We based our experiments on the assumption that an intent and concentrated thinking—affirmations, prayers, and the like—can program subtle energy to "do the job." In other words, we assumed that mental intent creates, in the subtle energy field, a pattern that carries the informational content of the thoughts. If so, this pattern can be imprinted by a healer into a substance, for which we used a water mineral solution. To create energy patterns, healers projected their intent for ten to fifteen minutes into a glass of the concentrated trace minerals water solution. This was done both locally, with the healer holding the glass in their hands, and from a distance, in which case the healer projected their intent to a glass at the VFT lab via Skype. Subsequently, the energy patterns imprinted by the healers were recorded and amplified using VFT equipment.

In an experiment described in chapter 7 and appendix E, Jones observed a dramatic increase in the survival rate of cells damaged by gamma radiation under the influence of pranic healers, as well as by subtle energy patterns produced by VFT. Among the VFT subtle energy patterns used was a particular pattern recorded at a distance from a healer. This pattern increased the survival rate of cells to the same percentage as the pattern of erbium, which is a phantom atom, used in the experiment. Two other subtle energy patterns used in these experiments were also subtle energy patterns created from a distance by the intent of an energy medicine healer and recorded by the VFT equipment. The intent of the healer was to protect cells from the negative effect of energetic pollution (creating the pattern called Protection), as well as to clean the lab environment from the pollution (creating the pattern called Clean Sweep).

The energy pattern Protection was infused into the cells' growth media. Clean Sweep was used in two different ways: in one experiment, it was transmitted into the lab several times a day through high-resolution audio equipment; while in another experiment this pattern was infused by the VFT equipment into water, which was regularly sprayed in the lab. These two subtle energy patterns were found to provide an increase in the survival rate of cells in a dirty (energetically polluted) lab from 0 to, respectively, 67 percent and 78 percent.

In another experiment, conducted with Dr. Jim Suiter, the subtle energy pattern Quantum Balance was recorded. This pattern targeted the support and strengthening of the muscle system of the body. Dr. Suiter had created an affirmation that—based on the testimonials of his patients—was very effective. We recorded the energy pattern corresponding to the affirmation. In this case, Dr. Suiter looked at a glass with a trace minerals water solution placed in front of him while projecting an image of a strong and perfectly balanced body. The recorded subtle energy pattern demonstrated the ability to noticeably enhance the muscle strength in people. Following are some testimonials provided by Dr. Tim Toula, who is a trainer of "extreme" athletes, on Quantum Balance's effect experienced by his trainees.

> Sunday, May 4th: Performed a bench press on a hammer strength plate loaded machine . . . the most I had ever done was 255 lbs. . . . After 8 drops of Quantum Balance in 6 ounces of water and 40 minutes later, I set a personal best of 280 lbs. of plates. . . . I consider this more significant than anything else I have done with Quantum Balance. I just kept adding plates until I had reached my max effort. . . . I could not believe that when I counted the plates, it totaled 280 lbs. . . . This was methodically done . . . and the weight increase is phenomenal . . . it was really hard to believe.

> Tuesday, April 29th: Performed a single-weighted pull-up with 140 lbs. of weight; my previous best had been 135 lbs. Thirty minutes prior to the maximum attempt, I drank 6 ounces of water with 8 drops of Quantum Balance mixed in. In my previous best attempt, I tweaked (i.e., strained) a few things in the process and had to go through a period of retraining for a max. After this attempt, I suffered no pain or "tweaks" or anything.

Dec. 8th, 2007: Little Rock City bouldering comp. Climber Ronnie Jenkins had been given a mixture of 8 ounces of water and 8 drops of Quantum Balance prior to the comp. One notable result was that he did "Bed Wetters," a V10 boulder problem, on his first try. He had never been able to do this particular problem prior to that day.

The above-described experiments open unprecedented possibilities for rigorous scientific research on many aspects of the mind-matter interaction, including experimental corroboration of our hypothesis that subtle energy is an instrument of this interaction. With the help of VFT technology, it becomes possible to experimentally demonstrate the ability of the human mind to program actions of subtle energy that can explain such phenomena as the power of prayers, laying on of hands, distance healing, and other paranormal effects. The same technology makes it possible to record the energy patterns that individual psychic healers use. Today, people often travel long distances to visit extraordinary healers. The same healing could be provided through recordings that can be infused into appropriate carriers, thus creating powerful healing tools for energy medicine.

Using the various methods we have presented, we have created a library of thousands of subtle energy patterns. These patterns are used to create sophisticated combinations that we call "VFT formulas." Since these formulas can be reproduced with VFT equipment, it opens up virtually limitless possibilities, ones *that never existed before,* for rigorous scientific research into the properties and effects of subtle energy. The ability to reproduce the same energy pattern, as many times as needed, allows us to satisfy all of the requirements of modern science regarding the stability and repeatability of experimental conditions. In upcoming chapters, we will discuss the results of experiments with VFT subtle energy formulas.

One of the main purposes of this discussion is to remove the veil of elusiveness from the image of subtle energy and show that its multifaceted effects on inanimate objects and living organisms confirm the reality of the title given to subtle energy by the ancients: the life force.

Human Instruments

From the beginning of our research on creating specifically targeted subtle energy patterns, we used several individuals possessing unusual abilities to see and feel and influence the subtle energy field. Along the way of our experiments with these medical intuitives, we developed protocols for testing various subtle energy patterns for their compatibility, in accordance with the purpose for which they were intended, and so on. Usually, we asked the same question to two or more of our human instruments and compared their answers.

One of these energy-sensitive people is Laura. I met Laura in 2012, when she came to me after my lecture about subtle energy and told me she was a psychic healer. At that time, we were experimenting with the recording of subtle energy patterns produced by the prayers practiced in different religions. It happened that I had with me a half-ounce bottle of a mineral solution infused with the energy pattern of the prayer Birkat Hamazon, used in Judaism. This prayer is to be spoken after finishing a meal. I gave the bottle, without any label, to Laura and asked her to describe what she felt. Laura took the bottle in her hand and concentrated for several seconds and then answered, "This energy helps very well with digestion. It spreads energy produced during the digestive process all over the body."

I was, as you can imagine, very impressed by the precise match of her description with the purpose of the prayer. It was one more confirmation that prayers and affirmations pronounced by humans are programming subtle energy. I suggested to Laura that we work together. In our first experiments, I showed her, via Skype, several bottles of solution infused by VFT with several variations of subtle energy patterns; all were aimed at the same goal. Laura was asked to "feel" which one was the best, and we showed her the bottles one by one. After the best energy pattern had been chosen, we changed numbers on the bottles and repeated the experiment, as a double-blind, again. Pretty soon, with more than a 90 percent success rate, Laura began to choose the same energy pattern as the best, again and again and again. Since then, for many years, we have performed hundreds of experiments of this kind, with a success rate close to 100 percent.

The following experiment may be an example of how human instruments can help with subtle energy research. Two sets of subtle energy patterns, created by VFT (six variations of Pattern A and three variations of Pattern B), were tested by Laura in several blind sessions, again via Skype, with the following results:

- Pattern A3 was chosen by Laura as the best among A1 through A6.
- Pattern B1 was chosen as the best among B1 through B3.

After that, another test was conducted to determine the optimal number of drops of the trace mineral solution* infused with the subtle energy patterns A3 and B1, which should be added to fifteen milliliters of some liquid supplement to make its effect on the human body as effective as possible. Several people were tested with this supplement in their hands, and Laura suggested seven drops of the solution with the subtle energy pattern A3 and three drops with the subtle energy pattern B1.

After that, we repeated the same testing with another energy-sensitive healer, Galina. To our astonishment, Galina chose as the best combination the *same* energy patterns—A3 and B1, and, even more astonishing, *precisely the same number of drops to be added to the liquid:* seven drops of the solution infused with the subtle energy pattern A3 and three drops of the solution infused with the subtle energy pattern B1. The odds of this occurring—taking into account that the number of drops of the energized solution needed to be added to the supplement could be up to one hundred drops (or about three milliliters)—are less than one in ten million.

---

*This supplement is called Omnimin TM Solution and is manufactured by Mineral Resources Int., Inc., based in Utah.

We conducted dozens of tests of a similar kind using Laura's and Galina's abilities, in order to sense various subtle energy patterns' effects on people, and we consistently found a great match with their choices (more than 90 percent). One can see how these supersensitive instruments, being combined with other testing methods including conventional medical tests on the biological effects of subtle energy patterns, can help to speed up subtle energy research and the development of its various practical applications.

Changes in Properties of the TMS Infused with VFT Formulas

We mostly use a TMS as a carrier of subtle energy patterns in our experiments. While other substances being infused with subtle energy patterns often change their effects, TMS precisely delivers a consistent biological effect of the specific subtle energy pattern to a subject. This has been confirmed by numerous experiments over the many years of our work with hundreds of health care practitioners in different countries.

We have found that only water containing some minerals can be used as long-term storage for subtle energy patterns. While mineralized water can hold subtle energy patterns for years, distilled water loses the subtle energy pattern infused into it in a matter of several weeks. In other words, the phenomenon of "memory of water," widely debated by many researchers (see chapter 7), depends on the presence of some minerals in the water.

The successful use of TMS as a carrier of subtle energy patterns raises an important question: What is changed in TMS by subtle energy patterns infused into it? Are these changes different for different patterns? Common sense tells us there should be specific changes in atoms contained in the TMS after infusion of certain subtle energy patterns, otherwise the TMS's energy pattern would remain the same.

If we accept a hypothesis about a connection between the subquarks' configuration in the nuclei of the periodic table of elements and their subtle energy patterns, we can argue that infusion might lead to variations in the configuration of UPA's in the nuclei of atoms that make up TMS. These variations would then result in supplementary variations of the subtle energy patterns of the solutions.

To observe changes in TMS under the influence of different subtle energy patterns, we conducted experimental research based on the assumption that changes in the subtle energy pattern of the trace minerals in water would be reflected in the structure of water. It is known that changes in the structure of water and other liquids can be analyzed using a gas discharge visualization (GDV) device.[15] That device is a computerized Kirlian camera that employs photographic images of electrical coronal discharges of objects placed in a strong alternating electrical field. When a drop of liquid (a trace minerals solution) is analyzed, the GDV calculates the number of fragments in the luminescent emission coming from the liquid. Not surprisingly, the number of fragments reflects the degree of the solution's structuring: a smaller number of fragments corresponds to a more structured solution.

Several energy patterns were chosen for the experiments. As a criterion for the difference between them, we considered their very different effects on physiological activities. The following subtle energy patterns were chosen:

1. This pattern has been shown to affect the brain's electrical activity.
2. This pattern demonstrated the ability to rejuvenate damaged cells (see chapter 7 and appendix E for more information on this).
3. Same as above, but this pattern adds the elemental quality of earth to the wood and fire qualities. The purpose was to see if changing an elemental quality of a pattern without changing its frequency spectrum would change the effect of this pattern on the trace minerals solution structure.
4. This pattern has demonstrated an effect on the autonomic nervous system (the pattern is called Stress Relief and is described in more detail in chapter 7 and appendix E).

4a. The same as subtle energy pattern 4, but this pattern has been infused into the trace minerals three times longer. The purpose was to see if a longer period of infusion would produce a noticeable change in the solution's structure.

The results of these experiments are presented in figure 4.3.

Figure 4.3 shows the GDV parameter called number of fragments (that is, fragments of Kirlian luminescence) as registered by the camera during several seconds of measurement. This parameter indicates to what degree the luminescent area (and, correspondingly, the solution) is structured. Again, the fewer the number of fragments, the more structured the solution is. Each measurement was conducted three times, to ensure consistency of the results.

One can see that the subtle energy patterns used in this experiment produce very different structuring effects on the solution into which they are infused. The number of fragments in the case of the mineral solution infused with subtle energy pattern 1 is practically no different from the un-infused solution, producing approximately 340 fragments.

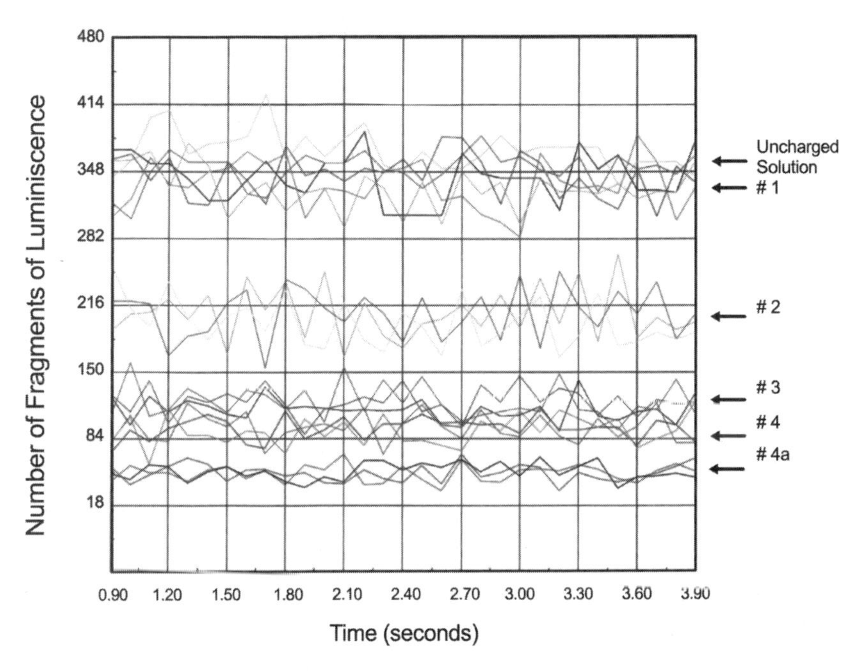

Fig. 4.3. This graph shows the results of a set of experiments that evaluated the effectiveness of several subtle energy patterns.

However, subtle energy pattern 4 produces a very strong structuring effect, producing approximately 56 fragments.

All other subtle energy patterns produced a structuring effect that falls between these two numbers. Furthermore, it is important to note that the curves for the various subtle energy patterns are clearly distinguishable from each other. It should also be emphasized that subtle energy pattern 3, with the elemental qualities of wood, fire, and earth, clearly has a different effect than pattern 2, which has the elemental qualities of just wood and fire. This indicates noticeably different properties of the pattern with the same frequency spectrum, but with different qualities of traditional Chinese medicine's five elements. This finding confirms what follows from traditional Chinese medicine: properties of subtle energy (or chi) cannot be adequately described by the amplitude-frequency language currently used in electromagnetic energy field theory.

Comparison of the curves for patterns 4 and 4a shows that infusion of the same pattern into a TMS three times longer caused a step-down of the fragments, as counted by the GDV, but the reduction was less than three times (in fact, it was one and a half times). This result indicates the presence of a saturation effect, when the substance is subjected to the strong subtle energy emission for a longer period of time.

Important conclusions can be derived from these experiments, when we take into account that changes in the structure of water indicate that the *interaction* of water molecules is changed as a result of the infusion of subtle energy patterns. This is possible only if the *configuration* of water molecules is changed. The molecules of water, H_2O, geometrically form a triangle (see fig. 4.4a) with an angle of about 105 degrees. Changing the interaction between molecules, which is necessary for altering the water structure, can occur only if the H_2O triangle is changed (see fig. 4.4b). This, in turn, is evidence that the interatomic bonds between hydrogen and oxygen atoms have been changed as a result of the infusion of subtle energy patterns.

As we discussed in the previous chapter, these alterations can be a result of structural changes in the configuration of the UPAs or subquarks in the nuclei of hydrogen and oxygen. Such a hypothesis seems logical.

Another confirmation of the changes happening in the water molecule, under the influence of subtle energy, is the experimental observation of changes in the frequency spectrum of vibrations of the molecule. These changes were detected with the help of Raman spectroscopy. This is a laser-based form of spectroscopy that measures a shift in the frequency of the scattered laser light occurring due to the vibrations of the water molecules. The intensity and frequency of the scattered light allow for determining the parameters of chemical bonds in the molecule. Changes in the Raman spectrum of water under the influence of subtle energy were first observed in the previously mentioned experiments with Dr. Xin.[5,16]

Subtle energy ("external qi," using the language of Chinese researchers), which was directed by Dr. Xin, dramatically changed the parameters of the Raman spectrum in various liquids: water, as well as saline, glucose, and medemycin solutions. Examples of changes in the Raman spectrum of water are presented in figure 4.4.[16]

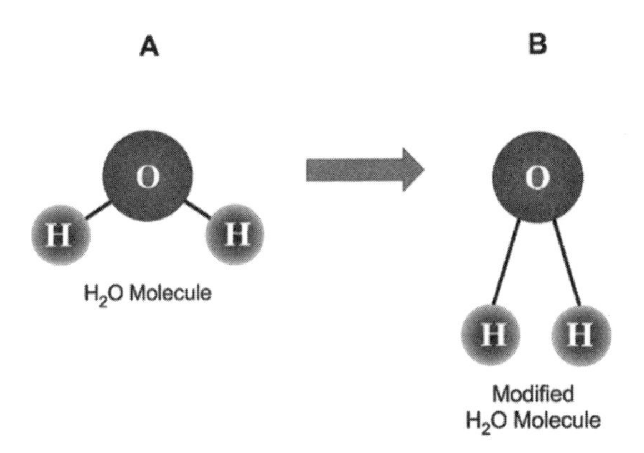

Fig. 4.4. Changes in the structure of water as a result of subtle energy infusion. Fig. 4.4a shows an untreated water molecule with the standard structure. Fig. 4.4b shows the new structure of a molecule treated with subtle energy patterns resulting in an alteration of the properties of the bonds between the atoms.

We need to mention here that, just as in experiments conducted on the alteration of the radioactive decay rate, Dr. Xin was able to induce changes in the vibrational spectrum of molecules from a great distance (more than one thousand miles). Such a finding adds further confirmation about the nonlocal nature of the subtle energy field and its intrinsic connection with the human mind.

Like in the case of structural changes in water discussed above, changes in the vibrational spectrum of water detected by Raman spectroscopy indicate changes in the chemical bonding of the water molecules, which is possible only if the properties of hydrogen and oxygen atoms are changed.

Raman spectroscopy is widely used in conventional physics. That is why we turned to it in an attempt to confirm the results presented here, which were achieved with the help of the GDV method, a protocol not yet accepted by the scientific community at large. Samples of TMS infused with several subtle energy patterns generated by VFT were sent to the Biomaterials Research Initiative at Penn State University's Materials Research Laboratory.* Raman spectra of the samples confirmed that different subtle energy patterns infused by the VFT equipment into a TMS produced diverse changes in the vibrational spectrum of the water molecules (see fig. 4.5).

Below is a quote from one of the researchers from the Materials Research Lab, Tania M. Slawecki, Ph.D.: "Our chemical analyses confirmed Dr. Kronn's claims: all samples were chemically identical within reason. Structurally, however, the initial Raman spectroscopy data indicated that they were different: each had a unique structural signature corresponding to differing vibrational energy states."[17]

Thus, *both* the GDV method and Raman spectroscopy† showed that the external subtle energy flow, whether generated technologically

---

*The author of that initiative, Professor Rustom Roy, was widely known as an initiator of many frontier projects, among them the well-known Whole Person Healing.

†Changes of the vibrational frequencies of water molecules under the influence of a healer employing therapeutic touch technique was observed using infrared spectroscopy as well.[18]

or directed by the human mind, can significantly change the subtle energy pattern of a substance.

As discussed earlier, along with the change of the subtle energy pattern of a substance, the structure of the molecules of this substance can be altered as well. In such cases, if this substance is later involved in a chemical reaction, then parameters of this reaction, such as its speed or its outcome, can be changed in comparison with the same reaction involving the same substance that has *not* been influenced by subtle energy. This finding can explain experimental observations that have been made regarding subtle energy's effects on the speed of chemical reactions.[5,12]

Summarizing the above, we need to emphasize that substances that seemingly look chemically identical can produce a *distinctly different* biological effect on a human organism. The differences in the subtle energy patterns of substances can exhibit themselves in two ways: through their effect on the energetic system (for example, acupuncture meridians or chakras) and through changing the path of

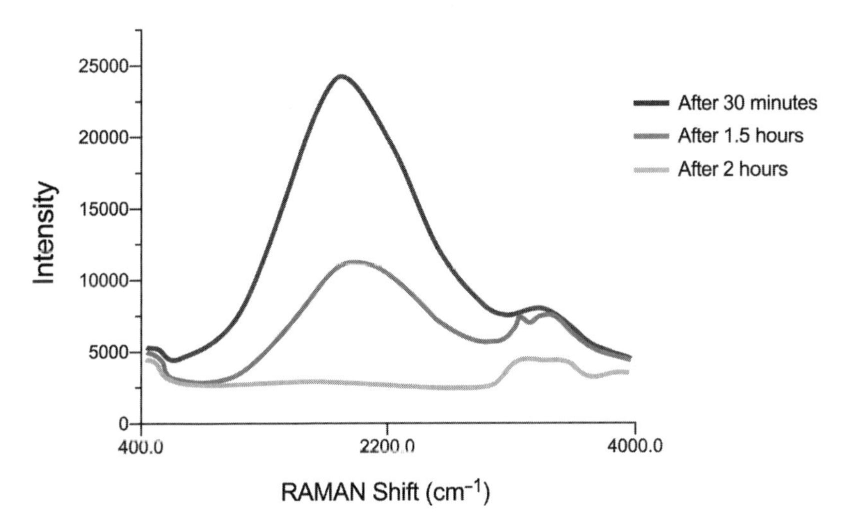

Fig. 4.5. This line graph shows the results of experiments on the effect of external qi, directed by Dr. Yan Xin, on the Raman spectrum of water.

the biochemical reactions in which the substance participates. Our experiments using TMS infused with different subtle energy patterns strongly support this conclusion.

The practical importance of this phenomenon depends, of course, on how long the energetically altered state of a substance lasts. As mentioned above, in our own rigorously controlled experiments with a concentrated mineral solution, specific biological properties of the subtle energy patterns (such as pattern 4, called Stress Relief, affecting the autonomic nervous system) could be observed *several years after the infusion*!

The results discussed above indicate that the lifetime of energetically altered water depends on the amount of minerals (that is, periodic table elements) that are diluted in the water. Organic, as well as crystalline substances, infused with subtle energy patterns using the VFT equipment have been shown to keep the energetic imprint for years. One can see these findings open up a wide range of possibilities for scientific research and practical applications of subtle energy.

Another practically important question is how efficiently substances infused with subtle energy patterns pass on their energetic properties to other substances, especially water, which is the most abundant common constituent of all living organisms on Earth. Undoubtedly, answering this question can provide keys for understanding the mechanisms of subtle energy's effects on living organisms, as well as show new ways of developing contemporary tools for energy medicine.

For instance, if the substance carrying a subtle energy pattern with specific energetic properties—healing or harmful—is able to effectively transfer its properties to bodily fluids, it is easy to see how the whole energetic system of the body can be affected. On the other hand, this phenomenon can also be used for delivering the *healing* properties of subtle energy patterns to living organisms. An experiment we will discuss next demonstrates that a mineral solution is one of the vehicles that can be successfully used as a subtle energy delivery tool.

Subtle Energy Transfer between Mineral Solution and Drinking Water

To investigate an example of the subtle energy exchange between substances, which has both scientific and practical significance, we used the GDV method to see how several drops of mineral solution infused with a specific subtle energy pattern could change the structure of regular drinking water. The results are presented in figure 4.6.

Water used in the experiment was drinking water, to which was added ten drops of energy-infused mineral solution per four ounces of water:

a. Water with un-infused minerals
b. Water with minerals infused with subtle energy pattern 4 (see the description of subtle energy patterns, given earlier)
c. Water sample directly infused by VFT equipment with subtle energy pattern 4

In the case presented, ten drops of the TMS infused with subtle energy pattern 4 (known as Stress Relief) added to four ounces of drinking water (a dilution of about 1:200) decreased the average number

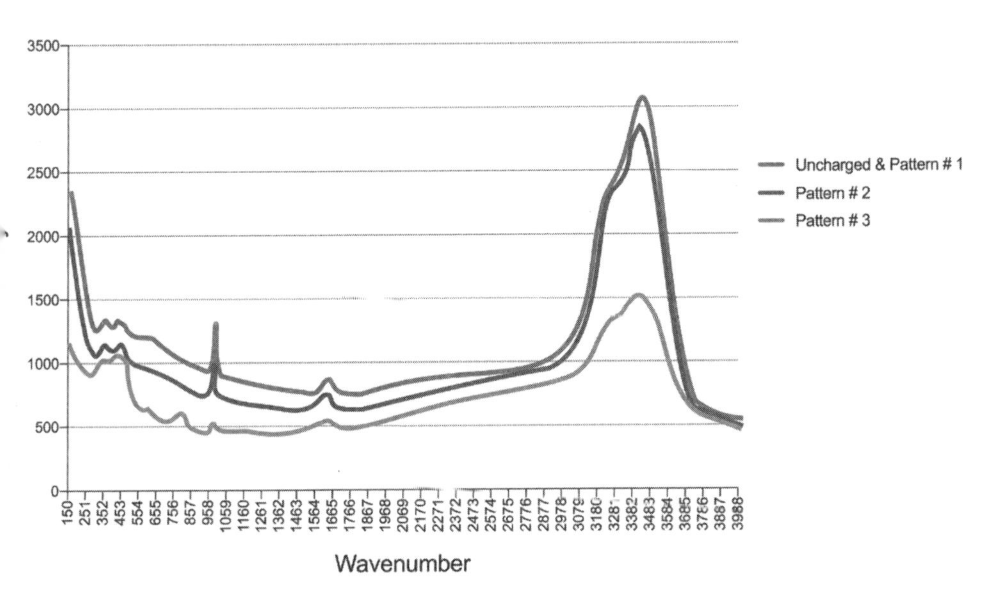

Fig. 4.6. This line graphs shows the Raman spectrum of mineral solution infused with different subtle energy patterns.

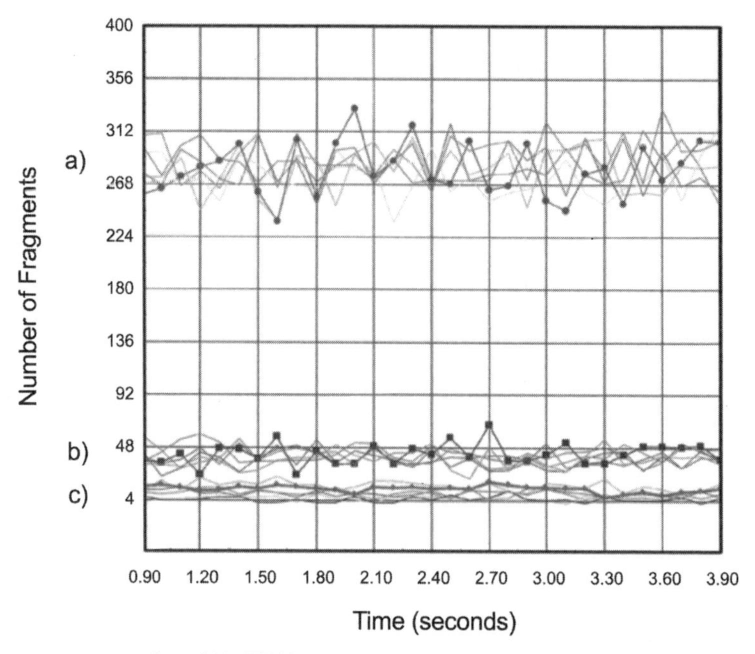

Fig. 4.7. GDV experiment: change in the number of fragments in drinking water associated with subtle energy transfer from a subtle energy infused mineral solution

of fragments in the luminescence of water from 280 to 40. One can see that the structuring of water created by these drops is close to the level of the structuring of water produced by direct infusion of the same subtle energy pattern (see fig. 4.7, curve c).

This indicates that just a small number of drops of the mineral solution infused with a specific subtle energy pattern can actually *transfer properties of that pattern* into a volume of water two hundred times larger. It appears the unique information of the pattern (including its distinct biological activity) was assigned to the whole volume of water. This can be reaffirmed by direct testing of the biological properties of water (see chapter 6 and appendix D for more on this).

We need to make some remarks about possible mechanisms of the biological effects produced by water structured differently than described above. Our experiments indicate that each subtle energy pattern creates a unique water structure. This suggests that just as any substance has a unique subtle energy pattern, so any water that has been structured (no matter how this structure was created) will have a subtle energy pattern specific for this structure. Thus, a conclusion can be proposed: The healing properties of structured water (if any) are defined by the subtle energy pattern specific to this structure.

We now know a mineral water solution can be an effective delivery vehicle for subtle energy patterns. It opens a wide range of possible practical applications and scientific investigations of the properties and effects of subtle energy. In the following chapters, we will present a series of experimental results achieved with the use of subtle energy–infused mineral solutions.

Conclusions

1. The process of the formation of the substances and energies of the physical world, following from Phillips's analysis of the observations of subquarks by Besant and Leadbeater, makes it possible to understand there are connecting points between electromagnetic and subtle energy. Consequently, this allows us to figure out technological methodologies to harness subtle energy.
2. Vital Force Technology (VFT), by providing a way to generate specific (as well as repeatable and consistent) subtle energy patterns, opens the door to rigorous scientific research into the properties and effects of subtle energy.
3. Experiments with VFT have confirmed that all elements of the periodic table, which represent the cornerstones of all known substances, are characterized by a unique subtle energy pattern.
4. The same frequency of vibration applied to plasma made of different inert gases generates subtle energy patterns with distinctly different properties. It confirms that mere frquency-amplitude language is not sufficient to describe subtle energy properties and supports the traditional Chinese medicine approach to subtle energy, as having five elemental properties (fire, earth, metal, water, and wood).

5. Experiments with water trace mineral solutions (a mixture of many of the periodic table elements) infused with distinctly different subtle energy patterns showed that the influence of a strong external subtle energy flow can dramatically change the subtle energy pattern of a solution. Also, these changes can last for years.
6. The ability of trace mineral solutions to hold the information encoded in subtle energy demonstrates the presence of an intricate and fundamental interaction between the subtle energy field and physical matter.
7. Subtle energy, generated technologically or directed by the human mind, can change the structure of molecules. As a result, it can influence the path of some chemical reactions (including biochemical ones).
8. "Structured" water healing properties, widely discussed by alternative health care professionals, are characterized by specific subtle energy patterns corresponding to particular structures. The lifetime of structured water's effects depends on the amount of mineral content in the water. Our research has shown demineralized water only holds the infused subtle energy pattern for a matter of days.
9. A relatively small volume of a mineral solution infused with a specific subtle energy pattern is able to transfer the properties (including biological activity) of the pattern to a volume of water hundreds of times larger. This finding makes it possible to practically apply this phenomenon for creating energetic remedies with specific biological effects. It also paves the way for research on the mechanism of homeopathic remedies' formation.

It needs to be stressed here that we are currently just in the beginning of building up a library of phenomenological scientific research on subtle energy. Many questions need to be answered to fully understand the wide-ranging rainbow of properties and practical applications of subtle energy. For instance, the results discussed in this chapter cannot explain why demineralized water is not capable of retaining subtle energy patterns infused into it. Does this mean that only periodic table elements heavier than oxygen (and therefore having a more complex nuclear structure) are capable of holding changes produced by an external subtle energy influence?

We hope future research into the nature of subtle energy will answer this and many other questions. Nevertheless, we will see in the following chapters that what we have already learned about subtle energy, combined with the ability to technologically generate subtle energy patterns, can rapidly propel forward our understanding of how to beneficially use subtle energy for scientific and practical purposes. It also has the potential to explain many puzzling supernatural phenomena, previously either ignored or denied by mainstream science.

5

Vital Seeds, Healthy Plants

SEED GERMINATION is a very convenient natural process for researching subtle energy's effects on living organisms. Unlike growing seeds in soil, standard seed germination research involves only water as the variable. Groundbreaking experiments were conducted by biophysicist W. C. Levengood (1925–2013).

In his experiments, Levengood used water infused with subtle energy patterns based on the emission spectrum of a single atom, in what he referred to as "phantom atom" energy patterns (recall chapter 4 for more on this). All experiments with germination were conducted using standard procedures prescribed by the American Society of Seed Testing.

The first experiments—conducted in 2002—were done with seven-year-old wheat seeds.[1] It is well established that older seeds have less vitality, creating a stricter condition for demonstrating the potential effects of subtle energy patterns. Springwater was used for the control seeds, while the experimental seeds were hydrated with water from the same source infused with various energy patterns of phantom atoms. The experiments employed energy patterns of various phantom atoms: silver (Ag), gold (Au), iron (Fe), lithium (Li), selenium (Se), and zinc (Zn).

The patterns demonstrated very different influences on germination and growth of wheat seedlings, ranging from suppression to very significant enhancement of germination rate and length, as well as seedling vitality (for a detailed description of the experiments dealt with in this chapter as well as summaries of the results, see appendix C). In other words, the results clearly indicated that different subtle energy patterns can produce dramatically different influences on internal biochemical processes that determine seed germination and further growth. Figure 5.1 presents photos of seedlings developed on the seventh day under the influence of the phantom atom of lithium, which turned out to have the most beneficial effect, in comparison with the control.

The water imprinted with the energy pattern of lithium stimulated a superior germination rate of 83 percent, compared to 43 percent in the control group. The seedlings growing in the lithium water have much more vigorous root systems and stem development.

Levengood also conducted an experiment with fresh wheat seeds. He wanted to determine if there would be a difference between the effects that the same energy pattern produces on old versus fresh seeds. Interestingly, Levengood found that water infused with the same energy pattern of the lithium phantom atom, which proved to be so beneficial for old seeds, did not help to improve the growth factor of the young seeds.

Assuming this outcome indicated an excess of the energy needed for the young seeds, he tested a set of progressive dilutions of the energy-infused water. Levengood discovered that the same results, as with the old seeds, could be achieved with the young seeds at a dilution of the

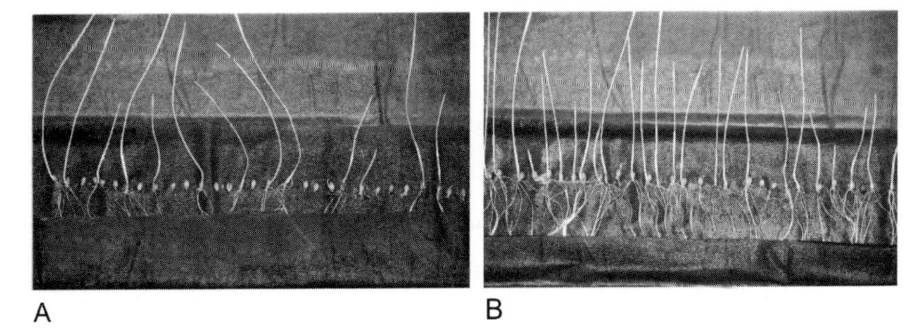

Fig. 5.1. The growth of seedlings on the seventh day in both a control group (A) and a test group (B) is shown in these images. Group B was grown with water infused with a lithium energy pattern.

energized water equal to 1:10,000. This finding opens the possibility for practical applications of subtle energy to increase crop growth and improve plant vitality. Indeed, there are a number of feasible ways of using modern agricultural watering systems to add a small percentage of energized water to benefit growing plants.

Levengood continued his experiments with phantom atom energy patterns using other plant seeds. He found that the maximal effect on the development factor of different seeds was achieved with different energy patterns. For example, for carrot seeds, the most effective energy pattern happened to be a pattern related to gold, while for pinto bean seedlings, it was a zinc phantom atom.

The experiments seem to illustrate that we need significantly different energy patterns to enhance the growth of different plant varieties. Though this deduction seems logical—and might be true in some cases—we need to take into account that the energy patterns used for the experiments presented here are simple ones: they only reflected one particular phantom atom. By comparison, innate plant growth is influenced by a great variety of energy patterns of chemical compounds in the soil.

Soil contains many minerals combined in complex molecules that have much more sophisticated subtle energy patterns than single periodic table elements. Taking this into account, we can assume that it may be possible to create sufficiently sophisticated subtle energy patterns capable of producing beneficial effects on a wide variety of plants.

The development of subtle energy patterns for practical applications in agriculture obviously demands long-term experiments in open fields. Research of this kind has not yet been undertaken at a large scale. Nevertheless, a yearlong experiment was conducted at Angel's Organic Farm, located in southern Oregon in the United States. At the farm, wheatgrass (used for juicing) was grown in a greenhouse environment using standard-size flats. Staff at Angel's Organic Farm conducted experiments using a subtle energy formula, developed with the help of VFT (recall chapter 4 for more details on VFT), to increase plant growth.

This "Healthy Plant" formula, as we called it, was composed from more than ten subtle energy patterns created with VFT. The patterns included energies of several composts and soil activators used in agriculture, subtle energy patterns of chlorophyll and humic acid, and a variety of phantom atoms, including those corresponding to phosphorus, sulfur, silicon, manganese, gold, and silver.

The greenhouse was divided in half: one side was set up as a control, and the other side was set up to test the Healthy Plant formula. The wheatgrass in both the control and experimental batches was grown using the farm's standard sprouting method. The only difference was in the water used. The control batch was watered using the farm's well water. The test batch used the same well water, but in this batch the water was also run through a filter containing man-made crystals that had been infused with the Healthy Plant formula. After two months of testing, it was clear that the wheatgrass grown with the Healthy Plant water was superior to the control batch.

Juice from wheatgrass grown with Healthy Plant water had a superior taste. The juice tasted sweeter but had a measurably lower sugar content. The treated wheatgrass grew 10 percent faster (see fig. 5.2) and

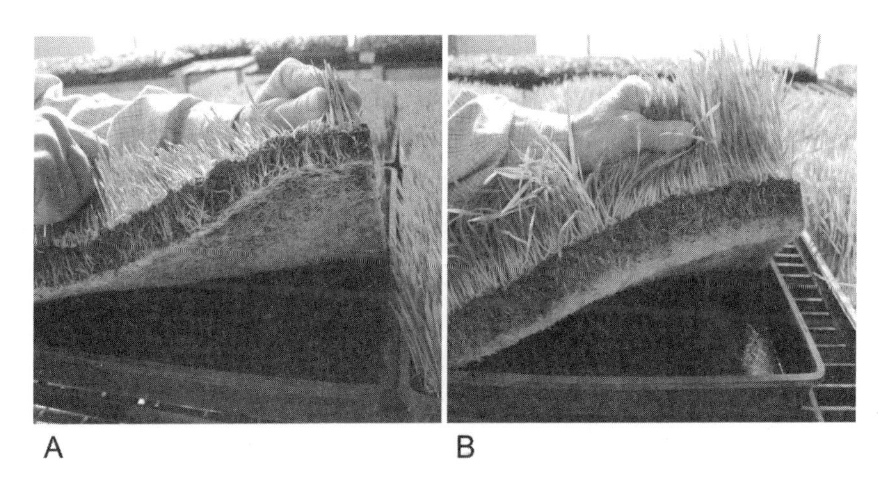

Fig. 5.2. These flats of plants were grown in an experiment that evaluated the effects of a subtle energy pattern called Healthy Plant. There is a noticeable difference in the roots of the wheatgrass grown in the control group (A), compared to the wheatgrass grown with the Healthy Plant energy pattern (B).

produced 60 percent more juice. Finally, the Healthy Plant wheatgrass had a longer shelf life.

Six years after the experiments at Angel's Organic Farm, the Healthy Plant formula was tested on onion and rice seeds. Keeping in mind the experience we gained working with Levengood, we used different dilutions for both the new and old seeds. In July 2015, experiments were conducted at the Seed Laboratory at Oregon State University in Corvallis, Oregon, where both seed germination and viability were tested. A batch of control seeds were watered with bottled un-infused water. Meanwhile, the experimental seeds were divided into groups and watered with infused water diluted in different proportions. In the lab, we observed that a dilution of 0.1 percent produced the best results for the germination of both rice and onion seeds. The treatment with the Healthy Plant formula increased the germination rate of both types of seeds, up to 164 percent for onion seeds and up to 193 percent for rice.

Again, several years later, in 2018, Angel's Organic Farm conducted a pilot study testing the influence of infusing a fertilizer with a subtle energy pattern on hemp. Two different types of hemp, Special Sauce and Sour Space Candy, were chosen for the trial. The fertilizer was infused with a formula that combined the energetic imprints of calcium, silicon, boron, magnesium, phosphorous, cobalt, and potassium.

The VFT formula showed high efficacy for Special Sauce hemp, yielding 84 percent more biomass over the control group of plants (see fig. 5.3 on page 105 for details). Research also demonstrated an increase in total yield, with the plants that received the VFT formula producing 14 percent more CBD per unit of weight.

These trial results show that a targeted approach, one that is focused on the development of a particular VFT energetic formula for a specific plant, may be beneficial both for enhancing the biomass of the plant as well as enhancing the chemical components of interest.

The observations made in the described experiments with plants allow us to make several conclusions that can be useful, especially when working with more advanced forms of living organisms, including the human body:

Results of the hemp growing with the infused fertilizer		
Plant	"Space Candy"	"Special Sauce"
	% over control	
Brix	14	7
Manganese	17	7
Zinc	29	17
CBD	7	14
Average Weight	0	84

Fig. 5.3. These Special Sauce hemp plants were grown in an experiment that evaluated the effect of infusing fertilizer with subtle energy. This image provides a comparison of plants grown in treated and untreated fertilizer.

- Different subtle energy patterns can have very different effects on the same plant, from strong support of germination and initial growth processes to significant suppressions of them. Therefore, we need to meticulously test the effects of all relevant energy patterns when pursuing the stimulation of any specific biological function.
- Old seeds demand a much higher amount of energy to stimulate their germination than young seeds. Obviously, this demonstrates the necessity of carefully determining the dosage of the energetic influence, based on the conditions of the seeds.
- Different plants demanded different subtle energy patterns for optimal stimulation of their development. This indicates that when trying to influence various biochemical processes, distinctly different energy patterns may need to be used.

- Complex energy patterns are able to produce more sophisticated changes in the biochemistry of plants than single phantom atoms. These patterns not only positively affected germination, growth, and number of plants but also significantly changed the chemical content of them.

These last observations lead to the conclusion that, the more complex biological processes we intend to influence, the more sophisticated the subtle energy patterns we likely would need to use.

For a detailed description of the experiments dealt with in this chapter as well as summaries of the results, see appendix C.

6

Energy Medicine

IF SUBTLE ENERGY—or the life force, as expressed by ancients throughout history—influences plants, we may expect that the same energy should affect all living organisms, including animals and humans. We should be able to notice this influence on all levels of the organism's structure and functioning: cells, organs, communications, and more. In other words, subtle energy should influence all operating systems of the living organism. There are—as of yet—no clinical studies about the effects of subtle energy on, for example, the autonomic nervous system, immune system, metabolism, or overall well-being of humans. But we cannot wait until the mainstream scientific community will recognize the necessity to do this important research. We want to move forward and advance our knowledge about subtle energy in the same way ancient practitioners did. For instance, Chinese acupuncturists collected, speaking in modern terms, case studies and summarized their experience. In this way, they developed algorithms for using acupuncture, including the principle of the five elements. Today, creative practitioners of energy medicine are doing the same, and many of them have achieved remarkable results using energy medicine tools separately or in combination with other healing modalities.

Dr. Jeffrey Marrongelle, a doctor of chiropractic and certified clinical nutritionist in the United States is one of the pioneers of such experimental case studies. For many years, he studied the effects that trace minerals infused with various subtle energy patterns produced on patients with different conditions. Marrongelle used the HRV test, defined earlier in this book, as a reliable scientific measurement of the autonomic nervous system response, and he tested several thousand patients in the course of his study. Measuring variability in heart rate serves as a dynamic window into the functioning of the synergistic action of the two branches of the autonomic nervous system (ANS), the sympathetic nervous system (SNS), and the parasympathetic nervous system (PNS). The ANS governs practically all physiological functions, including the action of the heart, secretions of different glands, and the assimilation and utilization of nutrients—it is like a computer running our physiology. Therefore, data about the condition of the ANS is an important indicator of the body's overall well-being. For example, an HRV test reflects positive or negative changes in a person's condition as a result of a treatment.

Based on his extensive experiments, Dr. Marrongelle concludes:

> Across the board and regardless of category or condition, consistent repeatable responses were observed. Positive shifts were observed in most of the heart rate variability parameters, particularly in the relationship of the sympathetic to parasympathetic reserve power. Vascular compliance, reserve capacity, and physical fitness measurements were seen consistently. Shifts in heart rate, myocardial response, high frequency and low frequency neurological function, "total power" and vascular compliance were observed repeatedly. It stands to reason, if one can provide positive influence to the autonomic nervous system as viewed by the HRV parameters, keeping in mind that the ANS is responsible for 99.99 percent of all life functions, the use of subtle energy to induce/initiate improvement of the ANS can significantly improve one's physiology and thus one's quality of life. . . . One of the major observations on the use of subtle energy patterns over time is their ability to enhance the biological actions of all other therapeutic modalities. In many cases once patients had a sufficient energetic influence induced and supplied by a specific energy pattern, they began to utilize homeopathic, herbal,

> nutritional remedies and pharmaceuticals in a more predictable and beneficial way. The data shows there was a decrease in paradoxical reactions to all forms of therapeutic intervention with the influence of these patterns.

In one set of Dr. Marrongelle's experiments, his patients were taking a water solution of trace minerals infused with the subtle energy formula Stress Relief. Subsequently, an HRV test showed that almost all parameters measured were significantly improved (for a detailed description of the experiments dealt with in this chapter, as well as summaries of the results, see appendix D). Among these, the parameter of optimum variability is an important one. An optimal level of variability of heart rate is critical to the inherent flexibility and adaptability that characterize healthy functioning systems. In a healthy individual, heart rate variability represents the effect of the PNS (it's the "rest and digest response," which slows heart rate) and the SNS (which is the "fight or flight response" we're all familiar with and which accelerates it).

The HRV tests performed by Dr. Marrongelle confirm the ability of subtle energy to influence all bodily functions, including the so-called supercomputer of the body, the ANS. The question arises: Which pattern of subtle energy should be used to improve a specific function (or functions) of a particular patient's body?

Dr. Marrongelle used electroacupuncture, created with the help of VFT, for testing the effect of different energy patterns on the endpoints of acupuncture meridians. Patterns were chosen that showed the best result in balancing and invigorating the meridian(s) connected to an organ (or organs) that had a problem. The energy patterns were infused into concentrated trace minerals, and several drops of these trace minerals were added to water that the patients drank.

Many practitioners of integrative medicine use electroacupuncture devices for testing various energy medicine remedies, such as homeopathy, flower essences, energy-infused substances, and the like. Advanced kinesiology is another modality that is used by practitioners for this purpose. Mainstream science can easily dismiss the outcome of these experiments as a result of the placebo effect, but subsequent experiments were conducted with mice using standard pharmacological methodology. The results of this study confirm the efficacy of subtle energy action in situations, with animals, where the placebo effect does not exist.

Pharmaceutical drugs are tested on animals to verify their effectiveness. While there is similarity of many physiological processes in animals and humans, the psychological functions can be significantly different. It raises questions about whether the same subtle energy patterns that are effective in humans are equally effective for animals, and vice versa.

A set of unique experiments was conducted on mice at the Laboratory of Behavioral Pharmacology, at University of Latvia in Riga, using the subtle energy VFT formula Stress Relief. In this research, a method called the Open Field Test was used. It is a validated method commonly used in neuroscience and pharmacology to measure changes in movement, such as exploratory and emotional behavior like risk assessment and anxiety-like behavior, in response to stress or remedies. It was the first time in the history of science that research about the effect of subtle energy on animals was conducted using standard pharmacological methodologies. The research was published in the peer-review scientific journal *Proceedings of the Latvian Academy of Sciences.*

In experiments with pharmaceutical drugs, animals are usually injected. During this experiment, mice were not injected; they were simply drinking water with trace minerals in it, and some of the water was infused with the Stress Relief energy pattern, while some was not. In the experiment, the mice were stressed with the artificially induced odor of a predator and/or they were forced to swim for several minutes, all while advanced software was used to track their movements.

The results of the experiment clearly showed that Stress Relief diminished the activity, or the stress, of the mice. The stressed mice who were drinking Stress Relief dramatically outperformed all control groups in the experiment, and the difference in activity levels with stressed mice not drinking Stress Relief exceeded 300 percent! This result shows that the Stress Relief energetic formula effectively normalizes the adaptive response to stress, which is necessary for safeguarding the body's health under the impact of stress. The experiment proved that the Stress Relief formula stimulates a strong anti-stress response and reduced the fear and stress of the treated animals.

Along with the Open Field Test, the blood glucose level in all mice groups was measured to determine physiological changes produced by stress. Measurements showed that stressed mice drinking the treated water had the same levels of glucose in their blood as non-stressed mice. However, stressed mice who were not drinking Stress Relief water had significantly elevated levels of glucose, another outcome that also supports the efficacy of the Stress Relief formula.

The results of the scientific experiments at the University of Latvia confirm practical observations of subtle energy effects on animals made by veterinarian Dr. Gary Tran at the Animal Emergency Center in Louisville, Kentucky. Over a period of a decade, he used various subtle energy patterns created by VFT on animals he was treating for different disorders. Describing his results, Dr. Tran says, "The energy patterns speed up the recovery of many severe and sometimes seemingly hopeless conditions. Reversal of severe symptoms like coma, intractable pain, shock, and hemorrhage occurs in hours, instead of days, like before." Some of the subtle energy patterns, according to Dr. Tran, "are very useful before and after risky and prolonged surgical procedures in traumatized and debilitated patients. Uneventful and rapid healing can be expected."

In summarizing we may conclude that properly arranged subtle energy patterns can significantly influence the health and behavior of animals. Thus, all experimental results confirm once again that subtle energy indeed has earned the title of "life force."

In our next series of experiments, we observed the subtle energy effects on human dermal fibroblast (HDF) cells. This research was conducted at Beech Tree Labs in Rhode Island and at the Department of Pharmacology at the University of Latvia. The experiment conducted at Beech Tree Labs was aimed at observing the effects of various energy patterns created with VFT on the viability and longevity of HDF cells. The control cells were grown in a standard growth medium, while the experimental cells were grown in the same medium that was infused with one of the following energy patterns: Cell Longevity, Peak Performance, Stress Relief, and Formula AC.

The experiments showed that the growth rate of cells grown in all of the energy-infused media was significantly higher than in the control group. The best result—82 percent more than the control—was achieved by the Cell Longevity energy pattern that was created by VFT specifically for the support of the cells' functioning. The Peak Performance energy pattern, known for enhancement of physical endurance, also significantly stimulated the cells' growth and viability.

The results also showed that cells in the energy-infused media died much slower than the control cells. This preserving effect of the subtle energy patterns on the cells situated in the toxic environment demonstrates that some subtle energy patterns are capable to act on all levels of the living organism—from cellular to systemic.

A next set of experiments was done at the Department of Pharmacology at the University of Latvia by Linda Klimavičiusa, Ph.D. She tested subtle energy's ability to influence a cell's viability and the mitochondrial membrane potential under conditions of food deprivation.[1] In addition, the researcher studied subtle energy's effects on cells when potent mitochondrial toxins are present.

The human embryonic kidney cell line (HEK-293) was used for this experiment. The cells were grown in a medium that was infused with Cell Longevity or Peak Performance subtle energy patterns. In the control group, a regular medium was used, and after several days, cell viability was tested using a spectrometer. The cell viability of the control cultures was taken as 100 percent.

The Cell Longevity energy pattern, like in the previous experiment, produced the best effect on cell viability: 46 percent higher than in the control, and the Peak Performance pattern showed a 21 percent increase in comparison with the control. Subsequently, when small amounts of mitochondrial toxins were added, Cell Longevity improved the viability of cells more than in the non-poisoned control.

Summarizing, we may conclude that all the experiments conducted in two separate laboratories demonstrate that subtle energy patterns targeted for the support of cellular health show an extraordinary ability to provide high viability and proliferation of cells in food deprivation conditions.

Experiments on Genes

Thirty years ago, a new branch of molecular genetics appeared in the biological sciences: epigenetics. Wikipedia describes epigenetics as follows:

> Epigenetics is the study of changes in gene expression caused by certain base pairs in DNA, or RNA, being turned off or turned on through chemical reactions. In biology, and specifically genetics, epigenetics is mostly the study of heritable changes that are *not* caused by changes in the DNA sequence; to a lesser extent, epigenetics also describes the study of stable, long-term alterations in the transcriptional potential of a cell that are not necessarily heritable. Unlike simple genetics based on changes to the DNA sequence (the genotype), the changes in gene expression or cellular phenotype of epigenetics have other causes, thus use of the term *epi*—(in Greek, *επί-* means "over, outside of, around") and *genetics.*

All biologists involved in the research of epigenetics agree the environment plays a decisive role in changing the expressions of genes. Some of them think an environmental effect has more influence on the cell's life than the classical genetic changes in the DNA sequence. In his book *The Biology of Belief* Bruce Lipton, Ph.D., states that "a cell's life is controlled by the physical and energetic environment and *not* by its genes. Genes are simply molecular blueprints used in the construction of cells, tissues and organs. The environment serves as a 'contractor' who reads and engages those genetic blueprints and is ultimately responsible for the character of a cell's life. It is a single cell's 'awareness' of the environment, not its genes, that sets into motion the mechanisms of life."[2]

Lipton argues that the "physical and energetic environment" influences the gene's activity. Can it be that the subtle energetic environment plays an essential role in the phenomenon of epigenetics? While many frontier researchers, including Lipton, believe this to be true, there has been no experimental proof to substantiate this belief, up until recently. In 2013, a first study was done at Beech Tree Labs in Providence, Rhode Island.[3]

Experiments were conducted in vitro using human dermal fibroblasts cells. The experimental cells were grown for twenty-four hours in a medium infused with a specific subtle energy pattern created with the help of VFT. Then, the activity of the cell's genes was measured using a sophisticated technique utilizing microarrays tailored to a specific disease or a cell process (like inflammation or immune response).

The results showed that several genes responsible for the body's reaction to pain were down-regulated. One of these genes—tachykinin receptor 1, TACR1—was down-regulated significantly. TACR1 is the gene that encodes for the receptor neurokinin 1. When stimulated by being bound with its activator, TACR1 is associated with the transmission of stress signals and pain, as well as the contraction of smooth muscles and inflammation. Down-regulating this gene would lead to less receptors being produced and would potentially result in a diminished response to the stimulating pain, decreased inflammation, and less stress on the body. Receptor neurokinin 1 can be found in both the central and peripheral nervous systems.

The next experiment conducted at Beech Tree Labs answered an important question: Can subtle energy enhance drug effectiveness? The experiment was done with a drug that targets and reduces myofascial inflammation, as demonstrated by testing numbers of inflammation markers. The subtle energy pattern called Anti Aging, which intends to slow down the aging of the human body's cells, was added with the drug to the growth medium of cells. The pattern was not added to the growth medium for the control group. The Anti Aging formula enhanced the drug's effect on the targeted genes regulating the immune system, between a difference in reduction of inflammation markers around 60 and more than 130 percent. Additionally, subtle energy significantly (more than threefold) up-regulated genes that the drug alone did not.

An additional experiment conducted at Beech Tree Labs demonstrated that a subtle energy pattern copied from a drug, called Adhesion-BTR and which was created at Beech Tree Labs, affected the same genes as the drug itself. In fact, the technologically enhanced energy pattern of the drug affected the same genes even more effectively.

Conclusions

- Subtle energy affects our life on all levels, including the deepest level of the regulation of biochemical processes in the body, the genetic level.
- The pilot experiments—focused on the regulation of drug activity by technologically produced subtle energy patterns—open a totally new perspective for the creation of tools for integrative medicine.
- The effect of subtle energy on the genes' activity can explain many puzzles of environmental influence on our lives, including the phenomenon of epigenetics.

These experiments are only a small step in that direction, and a huge amount of work needs to be done for the understanding of the subtle energy role in the area of genetics. Our hope is that conventional science will soon understand the reality of subtle energy's effects on our lives and will include subtle energy research in its agenda.

For a detailed description of the experiments dealt with in this chapter, as well as summaries of the results, see appendix D.

7

The Unseen Enemy

Energetic Pollution

THIS CHAPTER IS BASED on the experimental data presented by Joie Jones and Yury Kronn, Ph.D., at a health care professionals' convention in Baden-Baden, Germany, in October 2008, titled "Medical Week."[3,4]

Today, we know that pathogenic bacteria can cause many diseases. Dutch scientist Antonie van Leeuwenhoek discovered the existence of microorganisms in the 1670s, and he suggested they could be a cause of disease. However, it took two hundred years for medical science to accept this reality. It only happened in the 1870s, when British surgeon Joseph Lister developed practical methods for the sanitation of health care facilities.

Ultimately, the researchers Louis Pasteur and Robert Koch were able to persuade medical science that microorganisms could be a cause of disease and a source of contagion. In the two centuries between the discovery of Van Leeuwenhoek and the research of Pasteur and Koch, there were several scientists—among them Richard Bradley in the 1720s and Agostino Bassi in the 1810s—who presented serious experimental proof of the theory, but the medical community remained deaf to their voices.

This story illustrates how rigid dominating scientific paradigms can be.

The discovery we are facing today—subtle energy—is even more invisible than the harmful microorganisms that needed the invention of the microscope to be discovered and observed. However, the phenomenon is likely even more significant for our understanding of how environmental conditions can influence our health, the success or failure of various healing modalities, and the causes of diseases.

As we have seen, frontier science has provided rigorous experimental evidence of the existence of the effects of subtle energy on various strata of the physical world. Research has been done that demonstrates subtle energy's influences on human well-being and physiology.[1,2] Subtle energy, just like any energy in the universe, can have positive *or* negative effects. It is with this understanding that the extensive research of Professor Joie Jones, whose work we discussed in chapter 2, becomes highly relevant. Working at the University of California, Irvine, in the mid-1990s, Professor Jones began a long-term study to critically evaluate the effectiveness of the healing technique called pranic healing (also see chapter 2). For more than a decade, Jones used rigorous mainstream scientific methods to investigate possible meditation effects of pranic healing on human cells that had been subjected to gamma radiation. He chose human HeLa cells for the study, since radiation survival rates are well known for such cells, and the laboratory model he used is well established for radiation effects studies.

Jones describes pranic healing as follows:

> Pranic healing is a so-called biofield therapy established in China thousands of years ago, but rediscovered and reformulated in recent times by Master Mei Ling (sixth century AD) and Master Choa Kok Sui (twentieth century AD). Pranic healing is a type of "subtle energy" medicine in which the practitioners believe they are able to tap into what is termed "prana" or "chi," the "universal force," or "life force," and to use this "energy" to promote or enhance healing. Here "energy" is used metaphorically for processes we do not fully understand that seem to involve interactions above and beyond conventional energy.

The goal of the pranic healers involved in Jones's study was to alter the effects of radiation and enhance the survival rates of the cells. In his presentation to the Seventh Pranic Healers' Convention in Mumbai in 2006, Jones described the outcome of his experiment.

> In 520 experiments using 10 different pranic healers, typical survival rates increased from an expected 50 percent* for untreated cells to over 90 percent for cells treated both before and after radiation. The distance between the healer and the cells proved unimportant. Shielding the cells and the healer from electromagnetic radiation, including gamma rays, had no effect on the results. Four experiments involving groups of pranic healers suggest that group healing enhances the desired effect.

This research experimentally demonstrates the capability of subtle energy to reverse the detrimental physical damage to human cells. It also confirms the ability of the human mind to program subtle energy actions.[2]

Jones made another significant observation during the course of his research: he found that the healing effects produced by the pranic healers were dramatically different when conducted in energetically clean versus energetically dirty environments; in clean environments, the success rate was much higher.[3]

Jones's results mark the first scientific research with human cells that clearly demonstrates the damaging effect of energetic pollution on the healing process. It would be logical to assume that what applies to single cells also applies to a complex system of cells: the human body. In other words, the research of Jones indicates a sensitivity of the human body to energetic conditions in the environment. For example, an environment that is energetically dirty will tend to weaken a person's health and well-being (for example, due to the presence of chaotic energy originating

---

*HeLa cells subjected to the dosage of gamma radiation causing death of 50 percent of cells were taken as a control for the survival rate.

from various technological or environmental processes or residuals from interpersonal conflict or suffering). Conversely, the health and well-being of people will likely be more robust in energetically clean environments.

If this is true, we may also expect to see negative influences of energetic pollution on any healing modality or procedure. We should expect its effectiveness to diminish, slow down, or both. The results of Jones's experiments raise some important questions—with *huge* implications—in all areas of the health sciences. Certainly, medical clinics, surgical rooms, emergency rooms, psychiatric hospitals, and the like all contain energetic pollution. They all have, energetically speaking, something in common with the dirty lab where Jones conducted his experiments.

The possible presence of energetic pollution in potentially all health care environments leads to two questions: First, is it feasible to counteract the negative impact of energetic pollution on the healing process? And, second, is there any way to clean an energetically polluted environment? When I first met Professor Jones in 2001 at the Society for Scientific Exploration conference in La Jolla, California, we discussed these questions. I suggested that we set up a series of experiments using the subtle energy patterns technologically generated with VFT.

Jones understood the advantages of using technology in experiments with subtle energy to support both repeatability and standardized experimental conditions. He decided to test energetic patterns developed with the help of VFT. After several years of periodically repeated tests of VFT energy patterns, he made the following statement:

> I first encountered Dr. Kronn and his Vital Force Technology (VFT) in 2001 and was intrigued by his concepts and the VFT energetic patterns. Since then, I conducted a series of experiments with Vital Force formulas involving excitation of the Visual Cortex and the survival rates of HeLa cells subjected to gamma radiation. I found, using functional MRI, that when the acupoint BL67 was topically infused with a specific VFT formula excitation of the visual cortex

> occurred, similar to that produced by an acupuncture needle. These results suggest that the nature of subtle energy generated by VFT is similar to that of the energy referred to as "chi" in Traditional Chinese Medicine. Experiments with HeLa cells in culture subjected to gamma radiation showed that introducing cell compatible liquids (such as Ringer's solution) infused with VFT formulas specifically developed for cell rejuvenation increased the cell survival rate from an expected 50 to about 88 percent. These experiments suggest that Vital Force Technology could produce profound effects on biological systems and has enormous potential . . . to provide a firm scientific basis for Subtle Energy research and reveals new ways of integrating ancient Eastern Medicine with Western science.

In his first experiments with the VFT energetic patterns, Professor Jones determined the energy pattern that provided the biggest increase of the survival rate of the damaged HeLa cells (for a detailed description of the experiments dealt with in this chapter, as well as summaries of the results, see appendix E).

Just as it was found in the pranic experiments of Jones, the success rate—that is, the number of petri dishes with an increased survival rate—highly depended on the quality of the laboratory environment of the study. Energetically polluted environments clearly have a tremendous effect on the experiments. Energetic pollution literally obstructs the capability of the energy pattern–infused solution to increase the survival rate of cells subjected to gamma radiation. This fact was observed no matter how the healing energy was applied, whether by healers or through technological infusion. The fact that the subtle energy pattern generated by VFT equipment was much stronger than when projected by a pranic healer made no difference; it did not override the negative effects of the energetic pollution on the healing process.

These results inspired us to conduct additional experiments aimed at finding ways to protect the healing action of the subtle energy pattern infused into a Ringer's solution of several salts dissolved in water used for growing cells. In 2006, we set up an experiment aimed at finding a way to counteract the negative effect of energetic pollution in order to create conditions where the healing power of subtle energy could be observed consistently and in any laboratory. An additional goal of our experiments was to shed some light on the energetic mechanisms of the human mind's power of intent.

The energy patterns intended to fight energetic pollution were created by pranic healers and subsequently recorded and produced by VFT. Each pranic healer concentrated his or her power of intent for ten to fifteen minutes on a bottle of concentrated trace minerals solution in water. Energy patterns of this solution were copied, using VFT, thereby producing a new protective energy pattern. The presence of this energy pattern dramatically increased the number of successful experiments.

After the success of creating a protective energy pattern, we attempted to create a cleansing pattern that could energetically clean laboratories. Utilizing the same protocol we used for the creation of the protective subtle energy pattern, we made a cleansing energetic pattern.* In our experiment, after twenty minutes of energetic cleansing of the lab space, we observed an increase of successful experiments in a dirty lab from 0 percent to 68 percent. The results indicate that we may conclude that the frontier science technology of programming subtle energy patterns is capable of sufficiently counteracting the negative effects of energetic pollution.

Given the results obtained from these studies, we developed a spray infused with the cleansing subtle energy pattern as a simpler means of application to cleanse energetic pollution in a wide variety of spaces. The first clinical observations of the result of cleaning an energetic environment using the spray aptly called Clean Sweep were made by Marrongelle. He reported that, after spraying Clean Sweep around his patients, he observed positive changes in their ANS responses, which he measured using HRV testing.

---

*Later, energy patterns of several gemstones known for their cleansing abilities were added to this pattern.

In chronically ill people, who are typically overly sensitive to electromagnetic radiation, he observed a drop in the elevated heart rate of up to 20 beats per minute and a decrease in the sympathetic nervous system response up to 1.5 deviations toward normal. Even in people with no chronic conditions, Dr. Marrongelle observed that the spray had a calming effect on the sympathetic nervous system up to 1.5 deviations and improved a number of physical fitness parameters.

These results demonstrate the necessity for further research to see how cleansing an environment of energetic pollution will influence the results of various healing modalities. In recent years, positive clinical results have been reported regarding the use of this energy cleansing spray in business offices, veterinarian hospitals, and research labs in which animal experiments are regularly performed. These results indicate its efficacy as an energetic pollution cleanser. The findings of this research hold important implications for the study and practice of subtle energy medicine in a wide range of facilities, including hospitals, hotels, industrial and manufacturing plants, and factories.

Conclusions

- An energetically polluted environment significantly diminishes the healing effect of subtle energy.
- Special protective and cleansing energies are able to dramatically increase healing effects. This suggests a new method for improving the effectiveness of various healing modalities.

In closing out this part of the story, I would like to acknowledge the profound contribution of Jones toward the development of this new understanding of the environmental effects of subtle energy on the healing process. I believe future medical science will be comparing Jones's discoveries with Louis Pasteur's discovery of the role of germs and microbes in the causes and prevention of disease, which revolutionized health care science in the nineteenth century.

History of Water Memory Research

If the energetic environment affects healing modalities, it is also possible that the environment influences scientific experiments. This effect would be particularly evident with in vitro experiments, where biological objects are in close contact with the energetic environment.

As discussed in chapter 3, a significant contribution to the quality of the energetic environment comes from the experimenters themselves. A good example of this effect is an experiment conducted by French immunologist Jacques Benveniste that rocked the scientific world in 1988. In his article published in *Nature,* Dr. Benveniste and his colleagues reported that an allergy test had a positive result, even when the substance used in the chemical reaction (the reagent) was so diluted with water that no single molecule of the original reagent remained. As an explanation for the observed phenomenon, Dr. Benveniste suggested the water used for dilution essentially "remembered" information contained in the molecules of the original reagent.

If water is indeed capable of remembering information about the chemical properties of substances that have been diluted out of existence, it would open the door to confirming the efficacy of homeopathy. Soon after Dr. Benveniste's article was published and a heated discussion among scientists began, the editor in chief of *Nature,* Sir John Maddox, sent three experts to Dr. Benveniste's laboratory in Paris to repeat the experiment and see if the observed phenomenon could be reproduced. The composition of the team of experts illustrated the bias of the scientific community toward the possibility that water could have a memory. The team consisted of a physicist, a professional magician, and an expert in science fraud!

Not surprisingly, *Nature*'s team reported that they had failed to replicate Dr. Benveniste's group results. In 1989, the French National Institute of Health (INSERM) accused Dr. Benveniste of "insufficient critical analysis" of his results. Conventional scientists ridiculed Dr. Benveniste and even awarded him with the 1991 Ig Nobel Prize, a parody of the Nobel Prize.

Nevertheless, some scientists continued working on variations of water memory experiments to see if this phenomenon did have validity. A number of results turned out to be contradictory. While groups of Dutch (1992) and British (1993) researchers reported their failure to replicate Dr. Benveniste's results, a group with the well-known researcher Professor Madeleine Ennis in 2004 reported that, in three different types of experiments conducted in England and Europe, they observed a water memory phenomenon. Ennis, who began the research as a skeptic, stated, "The results compel me to suspend my disbelief. . . . We are however unable to explain our findings and are reporting them to encourage others to investigate this phenomenon." In 2010, Ennis, summarizing the results of twenty years of research on water memory, published a review in which she wrote, "Certainly, there appears to be some evidence for an effect—albeit small in some cases—with the high dilutions in several different laboratories using the flow cytometric methodologies. How much of the effect is due to artifacts, remains to be investigated."

She expressed her belief that in order to finish "the never-ending story" of the water memory research, there should be an agreement about the best way of conducting the experiment.

Based on the findings in this book, we intend to address both of these issues: how to conduct the experiments and how to explain the results. We will analyze the water memory phenomenon in the context of the energetic pollution of an environment.

In spite of the aforementioned damage to his reputation, Dr. Benveniste continued his research, and between 1997 and 2000 he published several articles in which he argued that the memory of water in a homeopathic solution is, by its nature, an electromagnetic signature. Accordingly, this signature could be sent over phone lines and the internet to a container of regular water, converting it into a homeopathic solution. The reaction of the conventional scientific community was to award Dr. Benveniste with his second Ig Nobel Prize in chemistry.

The next research into water memory was conducted in 2009 and 2010 by a group of scientists, including the French virologist Luc

Montagnier, who shared a Nobel Prize in medicine with two other scientists for the discovery of HIV. His group published a number of articles arguing that high aqueous dilutions of a viral DNA induced low-frequency electromagnetic waves. These waves are able to transfer the information contained in the DNA of the original organism to other ones. In an interview with *Science,* Montagnier said, "High dilutions of something are not nothing. They are water structures which mimic the original molecules." In other words, like Dr. Benveniste, he affirmed that water remembers properties of the original substance, even after diluting it so much that no molecule of the original substance is present in the solution.

An article by Andy Coghlan in *New Scientist* in 2011 described how scientists responded to Montagnier's findings: "A Nobel prizewinner is reporting that DNA can be generated from its teleported 'quantum imprint.'" The article continues: "A storm of skepticism has greeted experimental results emerging from the lab of a Nobel laureate which, if confirmed, would shake the foundations of several fields of science. 'If the results are correct,' says theoretical chemist Jeff Reimers of the University of Sydney, Australia, 'these would be the most significant experiments performed in the past 90 years, demanding re-evaluation of the whole conceptual framework of modern chemistry.'"

According to Mr. Coghlan, many researchers "reacted with disbelief," with Gary Schuster, former President of the Georgia Institute of Technology, comparing it to "pathological science." Biology professor P. Z. Myers also described it as "pathological science." He described the paper as "one of the more unprofessional write-ups I've ever run across."

This kind of power of paradigm response from the mainstream scientific community to experimental observations that are not explainable by current scientific doctrines is quite common throughout the history of modern science. Six hundred years ago, the vast majority of the population believed the Earth was flat. But for centuries prior to Columbus's proving the world is not flat, sailors and mariners had observed that a tall mast on a sailboat appears on the horizon before the ship itself does. Just think about it: this observation clearly proves that the Earth is not flat. Nonetheless, the flat Earth theory prevailed in spite of centuries' worth of evidence! No one discussed it; it was simply accepted as a fact: the Earth is flat. Period. It had become doctrine, and there was no room or need for discussion.

Today, we are facing a similar situation. In the past two decades, rigorous scientific evidence points to the existence of an extraordinarily significant, but not directly measurable, force and substance in the universe, which has been labeled dark energy and dark matter. According to scientifically accepted calculations and measurements, this force and substance make up approximately 96 percent of the known universe. It would be logical to expect that, after discovering this immensely powerful force that moves galaxies apart, scientists would start asking questions like:

- Is it possible that dark energy has other functions in the universe and influences other processes, including processes in living organisms?
- Is it possible other energies that were described and used in ancient medicines, like chi and prana, which also cannot be directly measured, somehow relate to this omnipresent, all-penetrating, universal dark energy?
- Can it be that some unexplained phenomena observed by science are related to the actions or influences of dark energy?

The mainstream scientific community is not addressing these questions because researchers are, by and large, either persuaded or coerced *away* from deepening their understanding about this immensely important force. Fortunately, there are a few prominent scientists who are in disagreement with the prevailing paradigm of conventional science. At a Nobel Laureates' Meeting in Lindau, Germany, in 2004, Brian Josephson, Ph.D., a Nobel Prize–winning physicist, gave a presentation aptly titled "Pathological Disbelief." He said, "The behavior of the scientific community . . . has a pathological component, in that the conclusions that were arrived at are, in all probability, different from those that would have been arrived at if the evidence available had been examined more objectively (had the committee members not taken a predominantly negative attitude to the evidence)." Josephson's observation is both problematic and troubling, especially for those either researching or practically applying frontier science. The credibility of any researcher—regardless of their scientific background—is called into question when he or she performs repeatable experimental research utilizing an energy that is unknown to science. The intellectual terror that came down on Nobel Prize winner Luke Montagnier is a good example.

The current scientific paradigm dictates what will be considered appropriate or inappropriate to study or worthy of the attention of science. British biologist Rupert Sheldrake writes in his book *Science Set Free:* "Contemporary science is based on the claim that all reality is material or physical. There is no reality but material reality. . . . [T]he belief system that governs conventional scientific thinking is an act of faith grounded in a nineteenth-century ideology. . . . Many scientists are unaware that materialism is an assumption."

Of course, when we choose pathways forward, we will be wise to acknowledge the prevailing paradigm . . . but *not* succumb to it. While we choose to apply rigorous scrutiny to our methods and conclusions, and diligently uphold time-tested standards with any practical applications, we must also stand true to our own experience, and our understandings—which *may* be informed by ancient traditions and modern science. That is, if both are viewed with an objective open-mindedness and regard for their respective approaches to the phenomena they're addressing.

From these principles, let us explore if we can explain any of the puzzles relating to the experiments on the memory of water. In chapter 4, it was shown that VFT makes it possible to copy subtle energy patterns of any substance and infuse those patterns into other substances, including trace mineral water solutions or even regular drinking water. We also established that the shelf life of infused trace mineral solutions lasts for many years. The oldest sample of TMS was infused fifteen years ago, and when it was tested in 2018 with the help of HRV measurement, it showed practically the same effect on the ANS as freshly infused solutions. Another demonstration of the long-lasting memory of a trace minerals water solution can be seen from the results of the experiments with the energy pattern Stress Relief, presented in chapter 6 and appendix D. A sample of the Stress Relief solution was sent to Latvian researchers, and during more than two years of research, it invariably demonstrated the same statistically significant effect on mice. In addition, the experiences of hundreds of health care practitioners in the United States and countries around the world using VFT energy formulas during the past twenty years provide confirmation of the existence and durability of the water memory effect.

As we have discussed in chapter 4, the ability to retain subtle energy patterns doesn't exclusively belong to water. Various subtle energy patterns, no matter how they were created—by a VFT generator, mapped from various substances, or created by human intent—can be infused into many organic and inorganic materials, including glass, crystals, oils, and a number of metals.

The remembrance of biochemical properties of various substances diluted in water, using a homeopathy method, can be explained by the interaction of water solutions with the subtle energy field (but not with the electromagnetic field).

We have a simple explanation and compelling evidence that the homeopathic remedies effect exists due to the subtle energy pattern of the substance that was diluted. Through many years of using VFT, we have copied the energy patterns of a substantial number of homeopathic remedies and infused them into trace mineral water solutions. The latter have *always* demonstrated the same type of improvement of physiological activity as the original homeopathic remedies. The only difference was they consistently demonstrated a *stronger* effect and were not as vulnerable to the environmental factors as the originals.

Many health care practitioners whom we've been working with for years have asked us to map homeopathic remedies—which we have done—and then gone on to use these trace mineral water solutions infused with homeopathic copies. They have reported back to us with great satisfaction, with comments like, "This method works for the enhancement of homeopathic effects like a rocket fuel."

We can conclude from the material presented above that the homeopathic method of remedy preparation is just one of the possible methods of copying subtle energy patterns of substances. Since this method is free from using electrical current, while many other methods are not, it produces very clean energy pattern copies. However, these copies are vulnerable, due to multiple dilutions. One explanation for their high vulnerability may be the effects of energetic environmental pollution, which has been discussed in this chapter.

Finally, we need to note that none of the scientists who experimented with the memory of water phenomenon during the past thirty years demonstrated they knew about the involvement of subtle energy in this phenomenon, nor did they know about the effects of environmental energy pollution.

Conclusions

- Energetic environmental pollution might affect not only the effectiveness of healing but also the outcome of scientific experiments, especially in vitro experiments when cells are directly exposed to the influence of the immediate environment.
- Using specific subtle energy patterns created for cleansing any dirty energy present within a laboratory's or research facility's space might significantly diminish the possible distorting effects of the pollution.
- The outcome of scientific experiments may be altered not only by technologically created environmental pollution but also by the energy of people participating in the experiment (for example, contravening or upset emotions, physically unhealthy people, and the like).
- Given that research in this area is in its infant stage, any new studies or experiments shedding light on this subject would be highly desirable.

Electromagnetic Pollution: A Growing Threat to All Life on Earth

Every day I get emails pointing out some new scientific research data on the damage to human health caused by electromagnetic radiation. Joel. M. Moskowitz, Ph.D., Director of the Center for Family and Community Health in the School of Public Health at the University of California, Berkeley, has created a website with thousands of research articles about this subject (www.saferemr.com). This accumulated research presents increasingly strong evidence of the multilayered damage that high-frequency electromagnetic radiation might cause to every living organism, including human beings.

A good example of the growing concern is the petition "International Appeal: Stop 5G on Earth and in Space," signed by more than 140,000 scientists, doctors, environmental organization representatives, and citizens from more than 200 nations. The petition summarizes the results of scientific research on the nonthermal effects of electromagnetic radiation on living organisms and states:

> In 2015, *215 scientists from 41 countries* communicated their alarm to the United Nations (UN) and World Health Organization (WHO). They stated that "numerous recent scientific publications have shown that EMF [electromagnetic fields] affects living organisms *at levels well below most international and national guidelines.*" More than 10,000 peer-reviewed scientific studies demonstrate harm to human health from radio frequency radiation. Effects include:
>
> - Alteration of heart rhythm
> - Impacts on general well-being
> - Altered gene expression
> - Increased free radicals
> - Altered metabolism
> - Learning and memory deficits
> - Altered stem cell development
> - Impaired sperm function and quality

- Cancers
- Miscarriage
- Cardiovascular disease
- Neurological damage
- Cognitive impairment
- Obesity and diabetes
- DNA damage
- Oxidative stress

Effects in children include autism, attention deficit hyperactivity disorder (ADHD), and asthma. Damage goes well beyond the human race, as there is abundant evidence of harm to diverse plants, wildlife, and laboratory animals, including:

- Ants
- Birds
- Forests
- Frogs
- Fruit flies
- Honey bees
- Insects
- Mammals
- Mice
- Plants
- Rats
- Trees

Negative microbiological effects have also been recorded. . . . [R]ecent evidence, including the latest studies on cell phone use and brain cancer risks, indicate that RF radiation is proven carcinogenic to humans and should now be classified as a "Group 1 carcinogen" along with tobacco smoke and asbestos.*

*All statements in this petition are substantiated by references to scientific articles published in peer-reviewed journals. Italics added by the author for emphasis.

An alarming aspect of the worldwide increase of electromagnetic pollution is the emergence of a new health problem: electromagnetic hypersensitivity (EHS). According to the petition, "EHS . . . affects an increasingly large portion of the population, estimated already at 100 million people worldwide, and that may soon affect everyone, if the worldwide rollout of 5G is permitted."

In her letter to California authorities, Beatrice A. Golomb, Ph.D., M.D., a professor at the University of California San Diego School of Medicine, who herself suffers from EHS, describes the potential danger of this disease spreading among the population.

> My research group at UC San Diego alone has received hundreds of communications from people who have developed serious health problems from electromagnetic radiation, following introduction of new technologies. Others, with whom I am in communication, have independently received hundreds of similar reports. Most likely, these are a tip of an iceberg of tens or perhaps hundreds of thousands of affected persons. As each new technology leading to further exposure to electromagnetic radiation is introduced—and particularly introduced in a fashion that prevents vulnerable individuals from avoiding it—a new group becomes sensitized to health effects. . . .
>
> Mechanisms by which health effects are exerted have been shown to include oxidative stress (the type of injury against which antioxidants protect . . .), damage to mitochondria (the energy-producing parts of cells), and damage to cell membranes, and via these mechanisms, an impaired "blood brain barrier" (the blood brain barrier defends the brain against introduction of foreign substances and toxins; additionally, disruption can lead to brain edema), constriction of blood vessels and impaired blood flow to the brain, and triggering of autoimmune reactions. Following a large exposure that depresses antioxidant defenses, magnifying vulnerability to future exposures, some persons no longer tolerate many other forms and intensities of electromagnetic radiation that previously caused them no problem, and that currently cause others no problem. . . .

> Affected individuals experience symptoms that cause them distress and suffering when they are exposed—symptoms like headaches, ringing ears, and chest pain from impaired blood flow; heart rhythm abnormalities; and inability to sleep. These symptoms arise from physiological injury. Moreover, many experience significant health problems that can include seizures, heart failure, hearing loss, and severe cognitive impairment. The mechanisms involved are those also involved in the development and progression of neurodegenerative conditions, including Alzheimer's disease.
>
> . . . [M]ost who are now affected were not, until they were. This may become you—or your child or grandchild. Moreover, if you have a child or a grandchild, his sperm or her eggs (all of which she will already have by the time she is a fetus in utero), will be affected by the oxidative stress damage created by the electromagnetic radiation, in a fashion that may affect your future generations irreparably.*

Dr. Golomb further suggests how to resolve this increasingly dangerous situation: "Let our focus be on safer, wired and well shielded technology—not more wireless." This would indeed be an ideal approach but, we cannot expect it to happen in the near future. However, we can think of possible ways to at least decrease the damage caused by electromagnetic pollution to our physiology. We have seen that a specifically formulated subtle energy pattern is capable of significantly increasing the survival rate of cells damaged by gamma radiation. This research result stimulated our scientific team to work on developing subtle energy patterns aimed at helping the human body to overcome the damaging effects of electromagnetic radiation. Our first attempt was directed to the development of an energy pattern helping the brain to function normally in the presence of the phone's radiation, because the brain is most influenced by eletromagnetic radiation during cell phone use.

*For the full document, see http://electromagnetichealth.org/wp-content/uploads/2017/10/Golomb-Beatrice-Sept-2017-FINAL1.pdf.

To determine the effect of our energy patterns, we used a well-known method of researching brain functions, the quantitative electroencephalography (qEEG), often called "brain mapping." This method is based on registering and analyzing electrical brain activity in the range of frequencies between 0 to 50 Hz. This range is subdivided into delta, theta, alpha, beta, and gamma frequencies, each of which is considered to be an indicator of specific neurophysiological states.

It seems logical to expect that research conducted by cell phone manufacturers, aimed at determining the effects of a cell phone's electromagnetic radiation on the brain, would include brain mapping. Strangely enough, you will not be able to find any publications coming from cell phone manufacturers containing the results of measuring the effects of cell phones on electrical brain activity. Standard tests of safety regarding a cell phone's effect on human health only include data of so-called SAR tests (specific absorption rate) that measure the amount of heat released under a cell phone's influence on the brain model, having electrical conductivity equal to the conductivity of human brain tissue. It seems clear it is advantageous for cell phone manufacturers to consider the brain simply as a piece of meat, instead of it being an immensely sophisticated biological computer with a delicate electro-neural network that can be detrimentally affected by a cell phone's radiation. With all of the evidence taken into account, it seems obvious external electromagnetic radiation could be disturbing to the brain's electrical activity.

After two years of experimenting with various energy patterns using the human instrument—people who are sensitive to electromagnetic radiation—we asked Jeffery Fannin, founder of the Center for Cognitive Enhancement in Glendale, Arizona, to do a pilot research study to see how smartphones influence electrical brain activity and whether specific subtle energy patterns might help the brain to maintain normal functioning in the presence of cell phone radiation. This pilot research was conducted with ten subjects using brain mapping equipment and software.

The experiment showed how cell phone communication causes excessive activity in a part of the frontal lobe of the brain that is associated with problems with working memory, such as spatial and visual, gestalt (configuring objects and experience), processing facial emotional expressions, and sustained attention. The elevation of neuronal activity in this region of the brain using a cell phone suggests that a person may have less efficiency in both emotional attention and verbal expression.

The tests also showed that the presence of the subtle energy pattern Transformer normalized the amplitudes of the brain waves in the areas that were previously overstimulated.

After analysis of all of the participants' brain maps, Fannin made the following conclusion:

> Results of independent tests . . . demonstrated higher levels of activity in all frequency bands (delta, theta, alpha and beta) from the left temporal regions continuous to frontal locations when cell phone use was engaged WITHOUT the infused "Transformer." . . . More normal brainwave activity is present in the areas examined when using the cell phone WITH the infused Transformer.

Clearly, these results point to *numerous* possibilities for using appropriately organized subtle energy patterns to counteract the negative effects produced by external electromagnetic radiation on electrical brain activity. Being encouraged by the demonstrated success of the first version of a subtle energy pattern designed for this purpose, we decided to investigate the possibility of making this pattern even more effective. As before, in our test research we used several people who had developed hypersensitivity to electromagnetic radiation.

To enhance the EMF Transformer energy pattern's effectiveness, we turned our attention to Schumann resonance. Dr. Igor Nazarov, who was a researcher at Energy Tools International for six years, describes this phenomenon in his 2019 paper, "Schumann Resonance, Brainwaves, Neuro-feedback and Beyond."

> [The Schumann resonance is] a natural phenomenon of standing, low-frequency electromagnetic waves manifesting themselves in the atmospheric gap between the surface of the Earth and the

> ionosphere. These waves come into existence mainly due to the intensive lightning activity in the atmosphere. Standing electromagnetic waves of Schumann resonance create a stable set of frequencies with an average base frequency observed at 7.8 Hz and harmonics at (approximately) 14 Hz, 20 Hz, 26 Hz, 32 Hz, etc. These rhythms correspond to the different brain waves—alpha, beta, gamma, delta, etc.—responsible for various mental states and functional characteristics of the human brain.[5]

Lightning radiates a very broad spectrum of electromagnetic waves, with frequencies ranging from several hertz to hundreds of megahertz. But the enormous spherical resonator formed by the Earth's surface and ionosphere only accumulates electromagnetic waves with wavelengths approximately equal to the circumference of the Earth, which is about forty thousand kilometers. Therefore, the frequency of these electromagnetic waves is about 8 Hz. (The frequency of an electromagnetic wave is equal to the speed of light divided by the wavelength.) The position and parameters of the ionosphere (including density, conductivity, etc.) are influenced by many factors, such as solar and cosmic radiation, day-night changes, and so on, and therefore are not very stable. This instability is reflected in the average deviation of the Schumann resonance frequency of about ± 0.5 Hz.

As in every resonator, along with the base frequency, its harmonics also satisfy the resonance principle. That is why electromagnetic waves with corresponding frequencies rounded to 14 Hz, 20 Hz, 26 Hz, 32 Hz, and higher are experimentally observed in the Schumann resonance phenomenon. Amplitudes of all harmonics are continuously fluctuating, but excluding rare occasions, they diminish when increasing the harmonic's number.

It has been determined that Schumann resonance frequencies reside inside the frequency ranges of electrical brain waves correlated with certain emotional and mental processes. Thus, alpha waves (7.5–12.5 Hz) correspond to the relaxed and calm state; beta waves (12–38 Hz) are mainly active during alertness and mental activity; and gamma waves, the subtlest in amplitude (38–100 Hz), are present during peak mental and physical performance. Some researchers even suggest that active gamma waves are related to expanded awareness and spiritual growth.

The relation between Schumann resonance and brain wave frequencies stimulated a discussion whether the everlasting presence of the Schumann electromagnetic wave could be essential for the development and functioning of all life on Earth, including the human organism. While there is no common explanation of how an external electromagnetic wave with such a long wavelength could influence the human brain, there is an indication that absence of it has a negative effect on human well-being.

Space agencies have discovered that the health of astronauts begins to decline after they spend months at a time on the space station orbiting the Earth. The Schumann wave exists only between the surface of the Earth and the ionosphere—far below the orbits of the space station. NASA subsequently reported that, after they'd artificially generated Schumann resonance frequencies (via an electronic device placed inside the space station), the health of the astronauts noticeably improved.

When searching for possible mechanisms of the Schumann resonance effect on human physiology, we need to realize that the ionosphere of our planet consists of charged particles—free electrons and the ionized atoms of atmospheric gases. Thus, the planet is surrounded by a thick layer of low-temperature plasma reflecting electromagnetic waves.[6] We know that gaseous plasma interacting with electromagnetic field vibrations creates a flow of subtle energy (see chapter 4). As a result, the Schumann resonance phenomenon also has, besides its electromagnetic component, a subtle energy effect. We need to take this into consideration when trying to understand how the Schumann resonance affects life on Earth.

We used a VFT generator to reproduce the subtle energy pattern corresponding to the main Schumann resonance frequency and its several harmonics. The plasma in the generator was modulated with these frequencies changing in the range of their deviations observed during experimental measurements of the Schumann resonance frequencies.[6] Then, the subtle energy patterns emitted by the generator were recorded to make them available for further research of their biological effects.

If the subtle energy component of the Schumann resonance is beneficial for the human body, we can expect it would diminish negative effects of any electromagnetic pollution. It would be reasonable to assume people who have developed electromagnetic hypersensitivity would be those who might notice it first. That is why we infused Schumann energy patterns into the EMF Transformer formula mentioned above (as an additive) and gave the new version of EMF Transformer to several people with EHS for testing. The reported responses of EHS people exceeded our expectations: they not only stopped experiencing painful symptoms they'd had before, but some of them even reported they felt better after using their cell phones with a modified EMF Transformer attached to it. Below are some examples of the testimonials we received.

> My entire adult life has been centered around cell-phones as my preferred method of communication. By the year 2000 and at the end of almost 20 years of heavy use of the cell-phones I was suffering severely from constant headaches & very specific sharp pain on my left-brain area.
>
> I became very sensitive to these EMF feelings and I had also searched heavily over the past years for any devices that would assist me with my problem. Once such device presented to me was the bio pro chip, which I immediately purchased and installed with absolutely no effects for me.
>
> I went to Energy Tools International for help. They provided me with an EMF sticker I placed on my phone. My bleeding in the ear stopped and I could finally tolerate up to 4 or 5 calls per day.
>
> After all the years and severe pain, this device was exceptional in every way. I could immediately deduct another enormous gain in relief to around 95% better than not using the device. I can now use

my phone in any emergency, for any length of time, with only minimal discomfort to the left side of my head.

I highly recommend this new unit to anyone who is EMF sensitive or anyone who cares about their health in any way. I personally believe it saved my life. —Ken B.

•◆•

When we were introduced to Energy Tools International [ETI], it was one of the blessings of life, because struggling to help protect people from EMF is a very difficult thing, since most of the products out there frankly just don't work. So, we spent a lot of time and money purchasing them for our clinic to evaluate devices. We went through years without finding much of anything.

We have tested many different products over the years and quite frankly, the only product that tests positive with the reduction of stress and geopathic stress has actually been ETI's product EMF Transformer. The EMF Transformer is so incredibly inexpensive by comparison of other things that are out there.

It is important for people to not only have these EMF Transformer protection devices on their cell-phones, but we also love the little pendants where you take the pendant infused with EMF Transformer and wear it right over the thymus. We have tested people where you have put the EMF Transformer over the thymus. After that, the body is so much less distracted and will maintain its focus on the healing and resolving of whatever conflict it is. My 15-year-old son is in a school that is bludgeoned by WiFi. We put an EMF Transformer on him like a necklace and found that it was incredibly effective to help him concentrate. We placed it right over the thymus.

The other thing we are seeing is—this is where the Energy Tools EMF devices came into play—we find a lot of people are sleeping with geopathic stress points. It is causing a profound weakness in their body, so they are not sleeping at night and they almost never find any peace either at night or during the day. We experimented with one of the EMF devices by folding it over. We had them place the folded EMF patch and put it under the pillow, and now they have no more exposure to geopathic stress. When we tested them later, it literally was resolved. —Michael R.

•◆•

I have found the addition of several of the Energy Tool products to an assortment of other treatments very beneficial. I find that Clean Sweep and the EMF Transformer almost fully eliminates my reaction to EMF, which has been an ongoing problem for years. —Joanne X.

•◆•

I used to feel a strange pain in my hand when holding my cell-phone. I purchased an EMF Transformer and applied it to my phone. I no longer feel pain when using my cell-phone. —Marie R.

•◆•

When I got the EMF Transformer yesterday . . . I did one muscle test with an iPhone calling and it was amazing to see the change of a weak test transform to strong with the application of the transformer to the phone. Wow! I am a believer . . . but then I've been a believer from the first time I ever used Vital Force Technology products." —Tim T.

Dr. Igor Nazarov stated in his review "Schumann Resonance and Beyond":[6]

"[B]ased on the testimonials of these people, it's clear these experiments may serve as evidence indicating Schumann resonance frequencies delivered as a subtle energy flow to the human body, especially the human brain, provide essential support for human well-being and might help people survive or even thrive in an electromagnetically polluted environment.

Further, it is hereby suggested that the subtle energy flow coming from the ionosphere, vibrating with Schumann resonance frequencies, is an essential natural neuro-feedback setup, which for millennia has been helping and guiding the human brain to evolve in harmony with our planet. Electromagnetic pollution surely makes it more and more difficult to 'hear' this subtle 'voice' coming from the ionosphere above, given it is covered by a cacophony of (electromagnetic) signals disorienting our brains. . . .

Of course, serious scientific research on the biological effects of subtle energy patterns corresponding to the Schumann resonance frequencies is necessary and could provide important information about the role of this phenomenon for life on our planet.

Conclusions

- There is growing evidence that continuously increasing electromagnetic pollution of the environment presents serious dangers to human health. The most evident and troubling sign of this is the increasingly large number of people reporting EHS, a mysterious new disease that damages various parameters of the physiology.
- The fact that electromagnetic radiation is influencing vibrations in the subtle energy field in general and thus might influence the energy flow in the human energetic system suggests this can be another way electromagnetic radiation can cause damage to the human physiology as well as to the physiology of all living organisms. Experimental research in this new area of science might shed light on the currently puzzling features of EHS.
- Successful pilot experiments with VFT subtle energy patterns aimed at counteracting negative effects of cell phone radiation

on the human brain indicate that using subtle energy technologies for the development of applicable tools could help diminish at least some of the damaging effects of electromagnetic pollution on living organisms. The positive influences of specific subtle energy patterns on plants, animals, cell growth, and health, as well as on gene expression (see chapters 5–7), also bear witness to the potential of such possibilities.

- The results of the research discussed above are an indication of the importance of the subtle energy constituents of Schumann resonance for living organisms. Further research of this phenomenon can give us substantial new information about both the mechanisms of Schumann resonance and the roles it plays in life on Earth.

For a detailed description of the experiments dealt with in this chapter, as well as summaries of the results, see appendix E.

CONCLUSION

A Tip of the Iceberg

IN THIS BOOK, we have described many experiments that prove the existence and influence of a force that—according to the current scientific paradigm—does not exist. We have seen that plants grow better when they get water that has "nothing" in it. We have seen that people who are too sensitive to electromagnetic radiation to use a cell phone can use such a phone with a "piece of plastic" attached to it. We have shown that it is possible to reduce the stress levels of animals, who are not susceptible to the placebo effect, with a treatment that modern science considers ineffective. These experiments were carried out rigorously and following accepted scientific standards. It is true that these experiments should be replicated at statistically significant sample sizes; however, we have enough compelling evidence for the existence and practical uses of subtle energy for any seriously open-minded scientist.

We may not yet know *how* subtle energy works, but we do know *that* it works. Remember, it is a fact that aspirin can take away a headache by inhibiting prostaglandins. We can observe its various effects, such as catalyzing enzymes, but still, we do not know exactly *how* aspirin causes these reactions in the body. That fact does not stand in the way of the widespread, undisputed acceptance of aspirin in society. The experiments we have described show that subtle energy influences life on all levels: systemic, cellular, and genetic. Subtle energy is truly a life force. And yet, it is a force that continuous to be denied. We are not talking about assumptions and hypotheses being ignored; we are talking about a lot of hard data being rejected. Some of the experiments we have described were replicated more than fifty times!

That ongoing denial brings up a major question at a critical moment in the history of humanity. In many ways, human civilization is at a crossroads with itself. There is no other species that, in an ongoing and massive way, exhibits behavior that directly undermines the very fundaments of its existence. We appear to be on a course toward self-destruction. I argue that we cannot progress as a civilization, even as a species, without the understanding and knowledge of subtle energy. And, importantly, most of this knowledge is not even new. It is merely a rediscovery of the wisdom all ancient traditions and cultures possessed.

I am not searching for a new law of physics that gives subtle energy a place in the current scientific paradigm. I am not looking for a brilliant equation that explains the mysteries of paranormal experiences. I do not even think that such a law or equation exists. Subtle energy interacts with consciousness. It is universally present, but it acts individually. It works not just with modern technology and equipment but with human instruments as well. To put it simply: it is impossible to explain the 96 percent of the universe that governs everything with a scientific mind-set that only understands 4 percent of reality.

It may very well require the emergence of a new species of humans with much higher levels of awareness to fully access the 96 percent. We need a new paradigm where spirituality and science can meet and where advanced meditators and groundbreaking physicists are the same people. In the meantime, however, we can investigate and use subtle energy enough to fundamentally change and improve our lives and society. We can move forward much faster and safer with more effect and without detrimental side effects.

"We cannot solve problems by using the same kind of thinking we used when we created them," said Albert Einstein. Such new thinking can begin with accepting the workings and influence of subtle energy in the same way we accept aspirin. The simple acceptance of the existence and influence of subtle energy opens the door to many practical applications.

When thinking about practical applications for subtle energy, three fields stand out: health, environmental pollution and food, and energy. Let us begin with addressing the opportunities subtle energy has to offer for health and healing. We know that Western medicine is good at critical interventions. There is no better system to repair a broken bone or to cure a life-threatening infection. However, today, in most of the world, most of the diseases are chronic conditions. In most of these cases, the best Western medicine has to offer is to treat the symptoms. But, true healing can only happen at a systemic level. Or, in the language of this book, compared to the treatment of symptoms with a pill in 4 percent of our reality, effective healing has to happen in the 96-percentage energy field.

Think about that: today, in every health decision doctors make, 96 percent of the information is missing. Another problem is that all the drugs of the pharmaceutical industry that do something good invariably also do something bad—the so-called side effects. The beauty of subtle energy medicine is that it never produces harmful side effects. The worst that can happen is that an energy does not work, because it is the wrong energy for that particular patient.

In the past twenty years, at Vital Force Technology, we have started building a comprehensive system of energy patterns for support of well-being on all levels: physiological, psychological, mental, and spiritual. In this book we have only presented the few patterns that have been scientifically tested. We have shown that these patterns are quickly and efficiently capable of decreasing stress—one of the biggest causes for any disease today—or reducing pain, helping the mitochondria to thrive even in a poisoned environment, or supporting the brain to function normally in the presence of electromagnetic radiation, and so on.

Through the years we have worked with health care professionals, medical intuitives, and meditators around the globe, and we have tested dozens of subtle energy formulas infused into trace minerals, oils, crystals, metals, cloth, and other materials. These formulas contain a wide variety of energy patterns, from gemstones and waterfalls to prayers and specific patterns directed by spiritual healers. Some of these patterns cover physiological issues like digestion and the immune system. Other patterns address physiological and mental issues like self-acceptance and forgiveness.

With more resources, our system of energy patterns for comprehensive well-being can be greatly expanded to include energy medicine tools for most conditions. The opportunity is massive. As we have shown in this book, a solution with the energy pattern of a molecule—without the actual molecule in it—can be as effective as the real thing. That means that we can create effective medicine without the need to deplete natural resources. We can record and reproduce energetic formulas that can be used forever at a fraction of the ridiculous costs of modern health care systems.

Today, people may travel to faraway places to visit extraordinary healers who are able to initiate miraculous healings. We can record the energies of these healers. That means that their healing powers can be made available to everyone without the need to travel. It also means that the energies of these healers, beyond their lives, can be preserved forever. Imagine fifty-five-gallon drums with a solution containing a particular energy pattern that can easily be shipped around the world to provide healing to millions of people wherever they need it.

Here is another example of the "out of the box" opportunities that arise when we accept the practical uses of subtle energy. Love may very well be the most powerful healing energy. Human beings can express and radiate love. The subtle energy of love can be captured and reproduced. The sand of the Middle East can be infused with the energies of peace and love. As far-fetched as that may sound, our experiments offer compelling evidence that such approaches and applications work. There are simple and subtle ways to reduce the anger, frustration, and stress in society.

A lot of the health problems in the world are caused by environmental pollution. We have polluted the lands with vast amounts of chemical fertilizers and pesticides. As a result, the food is polluted with chemicals and lacks essential nutrients. At the same time, the atmosphere has been polluted with ever-increasing electromagnetic radiation from cell phones and other modern technology.

Environmental pollution and radiation interfere with natural systems, including human bodies. The pollution disturbs ecosystems and energetic programs and makes humans more vulnerable to aggressive bacteria and viruses. Ultimately, health is a result of harmony in nature and all living systems. Through subtle energy applications, it is possible to restore harmony and improve systemic health. We have seen experimental proof of the influence of subtle energy on gene expression. This influence explains one of the major mechanisms of epigenetics, the study of biological mechanisms that turn genes on and off.

We have described experiments that show that subtle energy patterns improve germination and the growth of plants. More research in the areas of pollution and food will lead to even more opportunities to improve the health of people and planet.

Today, nuclear energy comes with the problem of radioactive waste. In chapter 2, we have discussed the experiments wherein Dr. Xin slowed down the decay rate of radioactive elements. Do these experiments show that it might be possible to clean up radioactive pollution? Theoretically, that should be possible. Subtle energy surrounds and influences all physical matter. We may never find such a revolutionary solution. However, one thing is clear: we will only find out if we spend our resources on ongoing research that begins with the acceptance of subtle energy as the essential life force of our universe.

More opportunities that today seem (highly) improbable will arise if we seriously turn our attention to subtle energy research. From time to time, claims of free energy devices keep showing up around the world. It is very likely that these experimenters and innovators have found ways to access the subtle energy field. Today, we cannot explain how the pyramids were built or how places like Stonehenge were constructed without the use of modern technology that was not available in those prehistoric times. The only reasonable explanation is that our ancestors were able to use the forces of nature in ways that we have since lost. It is the only reasonable explanation because we know that these ancient cultures referred to the life force—by whatever term they used—as the central force of their existence.

Sometimes, subtle energy is presented as a fifth force, after the four known forces in physics: gravitation, electromagnetism, the weak interaction, and the strong interaction. I think that description misrepresents reality. Those four known forces rule 4 percent of the universe. Subtle energy rules the remaining 96 percent. The term *fifth force* resonates with the unnecessary fifth wheel of a vehicle. The subtle energy field is not an *addition* to the known forces of physics. Subtle energy rules *at the core of our existence.* It influences and affects everything that involves the 4-percent forces.

There is a long road ahead of us. Our research, so far, touches only a tip of the 96-percent iceberg. Far more research needs to be done. More experiments, like the ones we have described, need to be done to find more practical applications to support health and well-being. We are looking for collaboration with all researchers, groups, and organizations interested in further research of the effects of environmental energetic pollution on living organisms, as well as the development of subtle energy–based tools to counteract the negative effects of environmental pollution.

Interestingly, given the very characteristics of subtle energy we have discovered, some of that research may turn out to be quite different from what we are used to. In that respect, I would like to refer once more to the groundbreaking research that was done by the two spiritual teachers Annie Besant and Charles Leadbeater in the early twentieth century (see chapter 3). Besant and Leadbeater used a meditation technique that today we would call remote viewing to precisely and accurately describe the elementary particles in the atom. It took billions of dollars and almost a century to build a machine—a particle accelerator—to confirm their findings.

The research of Besant and Leadbeater is far from easy. It requires spiritual discipline and rigorous training, but it does not require billions of dollars. The most important ingredient for groundbreaking research is the open scientific mind that the world of subtle energy, so far, so desperately lacks. I hope that the detailed descriptions of our research in this book will inspire scientists to join our mission to explore what we cannot yet explain and understand. At the same time, it is my wish that this book speaks to the layperson interested in finding better ways of healing. There is an opportunity in 96 percent of the space around us to make our world a healthier and safer place. It is a vast and exciting opportunity that we cannot afford to ignore.

APPENDIX A

Electrons, Subquarks, and Superstrings Examined by Stephen M. Phillips, Ph.D.

IN HIS BOOK Phillips analyzes the formation of the strong and weak forces shaping the physical world; but it would be valuable also to discuss the formation of another fundamental force of the atomic universe—the electromagnetic force. Here we intend to discuss, based on his material, the origin of the electrical charge, which is the source of the electromagnetic field.

After detailed analysis of the subquark structure of fifty-three "micro-PSI atoms"—described by Besant and Leadbeater—Phillips made a conclusion that a fixed electrical charge (positive or negative) cannot be assigned to either the positive or negative UPA. Here, we need to take into account the fact that it is now known to science: up and down quarks have fixed electrical charges, + and –, respectively.

As Phillips demonstrated, both of these quarks can be seen as Besant and Leadbeater's "UPA triplets," interacting with each other through the components of the force pouring through UPAs from the fourth dimension into the physical world; modern science calls this interaction colorful force. When looked at through scrutinizing eyes, one can see that, in the course of this interaction—a dynamic interplay that changes the properties (such as color) and behavior (or vibration) of the whorls making up the UPA an electrical charge is created. Thus, the logical supposition would be that different states of the superstring/UPA and their colorful interactions can create different values of electrical charge, as well as different signs (i.e., positive or negative).

Another support for this logical conclusion comes from Phillips's hypothesis about the nature of the electron. When Leadbeater looked for electrons emanating from the cathode of the vacuum tube, all he noticed was a flow of ordinary UPAs. Phillips commented on this observation.

> Leadbeater's expectation that electrons would look very different made him ignore the obvious possibility that the free "ordinary [UPAs]" were just what they have been looking for, namely electrons! If electrons resembling UPAs were indeed observed . . . (for the "ordinary [UPAs]" should be confined, not free, if they were subquarks), this would indicate that . . . the electron is a *single* particle whose 3-dimensional form (though not necessarily size) is sufficiently similar to that of a UPA to make Leadbeater believe mistakenly that he was observing free UPAs.
>
> This conclusion is consistent with the identification of the UPA as a subquark state of the superstring and with subquarks and leptons* being equally elementary, because the latter are also states of the superstring.[1]

If, as follows from the above, the electron and subquark/UPA are just different states of the superstring, then the hypothesis that different states of the superstring and interactions of superstrings create different electrical charges is true. In other words, we see that the electron state of a superstring has an electrical charge of –1, while the subquark states of a superstring, bound into up and down quarks, create charges of + and –, respectively. This fact can be explained only with the above suggested hypothesis.

One more time, we need to note here the fact that electromagnetic field acts only in that part of the world where an electrical charge is created—in our 3-D world—where it is created by interactions of the UPAs, as we discussed above.

*The electron belongs to the lepton family.

APPENDIX B

Fibonacci Sequences of the Traditional Chinese Medicine Elements

The He tu and Luo shu are two of the most famous diagrams in Chinese traditional culture as well as in the traditional Chinese medicine system.[1] According to legend, both He tu and Luo shu initially emerged as groups of black and white dots, where black dots represent even numbers and white odd numbers.[2] In particular, the group of dots with number five is positioned in the center of He tu and Luo shu and arranged as a cross. Further, according to the Hong fan section in Shang shu,[2] number one is an element of water; number two is fire; number three is wood; number four is metal; number five is earth. The number of different groups of dots in He tu or Luo shu can thus form a direct association with the five-element theory in traditional Chinese medicine.[1]

Taking as the basis two first numbers assigned to the particular element (four and nine for metal, for example, or one and six for water, etc.) we can use the Fibonacci sequence approach and calculate a series of numbers corresponding to each of the elements (see table B.1). Using these numbers as the modulating frequencies of the plasma generator (which was discussed in chapter 4), we were able to create subtle energy patterns that were tested on acupuncture meridians assigned to different elements. The result was that we successfully used EAV devices to influence the body the way it is described in traditional Chinese medicine. In other words, energy patterns created with frequencies related to the particular element had a stronger influence on the meridians associated with that element. It supports the supposition that the frequencies presented in table B.1 might be used for creating an energy pattern with specific elemental properties.

TABLE B.1

Metal	Water	Wood	Fire	Earth
4;9	1;6	3;8	2;7	5;10
92	86	79	66	65
149	139	128	107	105
241	225	207	173	170
390	364	335	280	275
631	589	542	453	445
1021	953	877	733	720
1652	1542	1419	1186	1165
2673	2495	2296	1919	1885
4325	4037	3715	3105	3050
6998	6532	6011	5024	4935
11323	10569	9726	8129	7985
18321	17101	15737	13153	12920

Table B.1. The numerical Fibonacci sequences associated with each element of traditional Chinese medicine. These were used to program the frequencies of the plasma generator.

APPENDIX C

Phantom Atom Effects on Plants

W. C. LEVENGOOD'S EXPERIMENTS with germination were conducted using standard procedures prescribed by the *American Society of Seed Testing*. Using special germination paper, up to thirty seeds per roll were hydrated and placed in a germination chamber with constant temperature. Germinated seeds were counted and the seedlings' heights were measured on the third, fifth, and seventh days of the experiment.

A summary of the results for the seventh day of the experiment is presented below in table C.1, using development factor, a parameter introduced by Levengood. (Development factor is the fraction of seeds that germinated multiplied by the average seedling height.)

TABLE C.1

Energy Pattern						
	Se	Fe	Zn	Ag	Au	Li
DF, %	-32	-3	50	56	58	77

Table C.1. This table gives a summary of the results that were achieved when W. C. Levengood experimented with seedling germination. The development factor (DF) is shown here, across several different phantom atom energy patterns, in comparison with a control group.

Figure C.1 shows that the difference between seedling development factors of the sample with the Li-PA energy pattern and the control was increasing day after day.

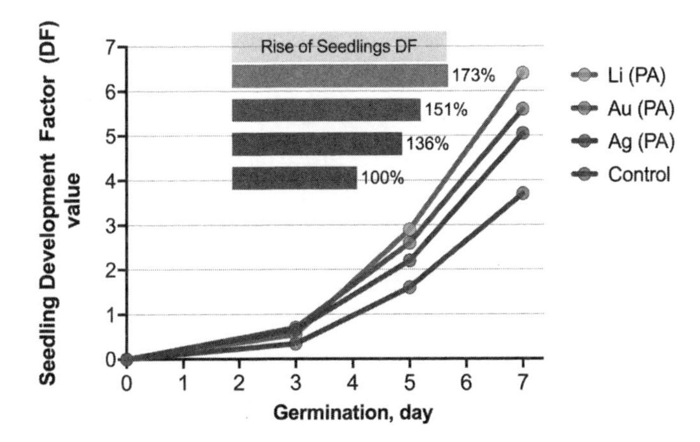

Fig. C.1. This chart compares the influence of beneficial energy patterns on seven-year-old wheat seed germination. The development factor, shown on the y-axis, is equivalent to the germination fraction multiplied by the average seedlings' height. The lines labeled Ag (PA), Au (PA), and Li (PA) represent silver, gold, and lithium phantom atom patterns, respectively.

Figure C.2 shows the results for progressive dilutions of the energy-infused water.

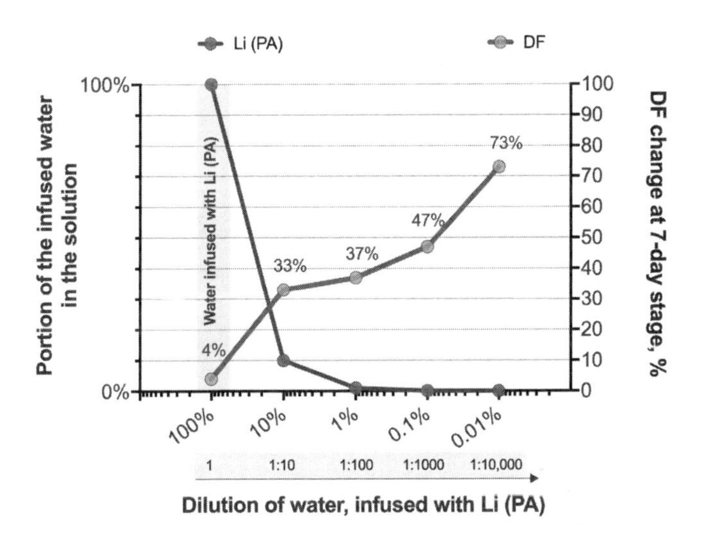

Fig. C.2. This chart shows the effect of different portions of infused water on the development factor of wheat seedlings, which were grown from young seeds in serial dilutions of water infused with a lithium energy pattern.

We can see that for young seeds the same results as with old seeds can be achieved at a dilution of the energized water equal to 1:10,000 (to see this, compare fig. C.2 with fig. C.3).

Figure C.3 presents photos of the pinto bean seedlings control group and a group germinated in water infused with a zinc energy pattern.

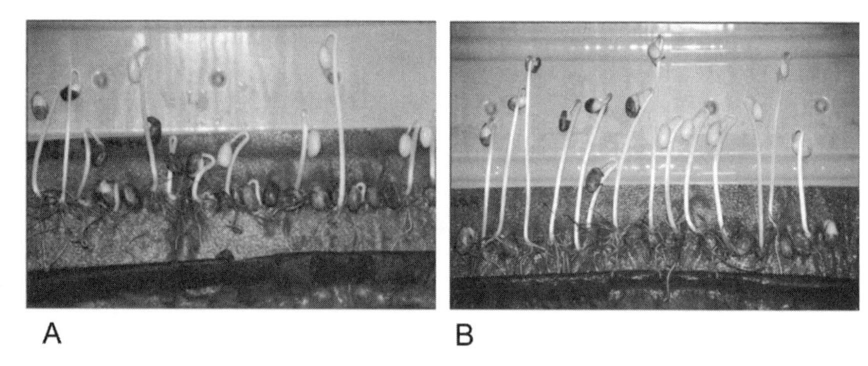

Fig. C.3. These images compare the three-day germination of seeds in a control group (A) and a test group grown with water that was infused with a zinc energy pattern (B).

One can see that, from the beginning of their development, pinto bean seedlings germinated in energized water are stronger than those of their control group. The difference between the seedlings' vigor becomes much more pronounced on the seventh day after germination (see fig. C.4). While less than 15 percent (2 from 14 germinated) of the control group reached 30 cm, 75 percent of seedlings in the experimental group (12 from 16 germinated)* reached 30 cm or more.

Figure C.5 shows the distribution of seedling heights in a control group and in experimental groups germinated in water infused with subtle lithium energy patterns—that was the most effective subtle energy pattern for wheat seedlings—and zinc. It is obvious that for pinto beans, the zinc phantom atom energy pattern produced the most robust seedlings. On average, they were 50 percent taller

*The total amount of seeds planted for germination was twenty in each group.

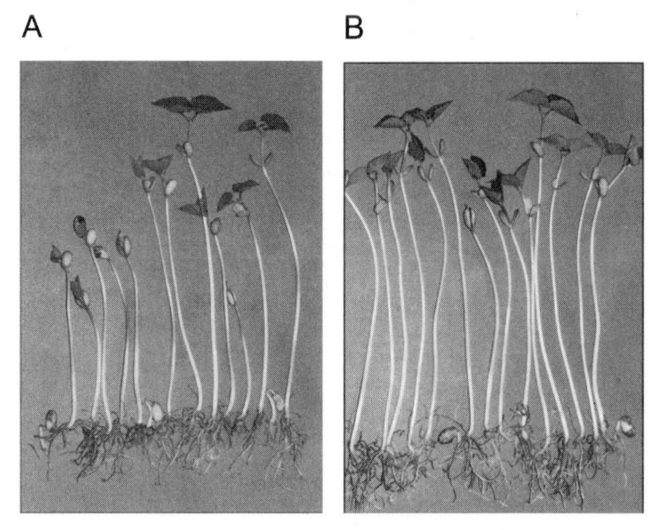

Fig. C.4. These images show the growth of pinto bean seedlings after seven days in control group (A) and test group that was grown with water infused with a zinc energy pattern (B).

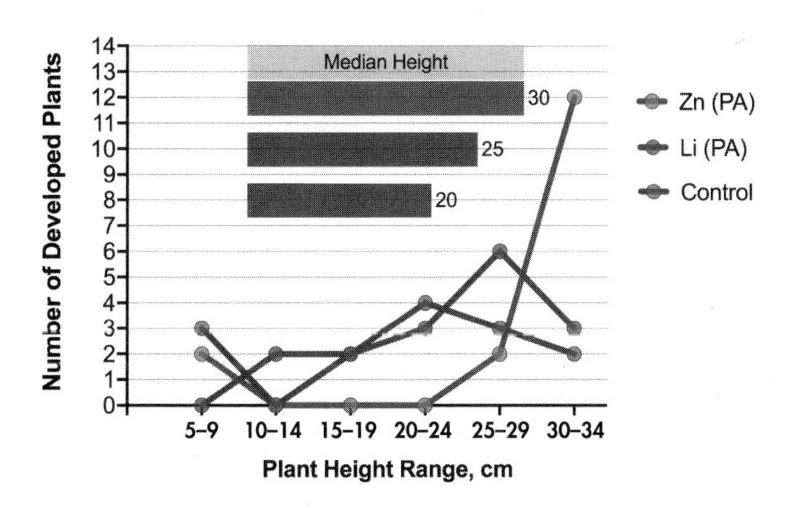

Fig. C.5. A height analysis was conducted on seven-day-old pinto bean seedlings, and this chart compares the results for a control group with two test groups using zinc, or Zn (PA), and lithium, or Li (PA), phantom atom patterns.

than those in the control group. The lithium phantom atom pattern was also beneficial for pinto bean development, producing on average seedlings that were more than 20 percent taller than those in the control group.

Experiments with the Healthy Plant Formula at Angel's Organic Farm

The greenhouse in the experiments at Angel's Organic Farm was divided in half: one side was set up as a control, and the other side was set up to test the Healthy Plant formula. The wheatgrass in both the control and experimental batches was grown using the farm's standard sprouting method. The only difference was in the water used. The control batch was watered using the farm's well water. The test batch used the same well water, but it was also run through a filter containing man-made crystals that had been infused with the Healthy Plant formula. After two months of testing, it was clear that the wheatgrass grown with the Healthy Plant water was superior in comparison to the control.

After a year of observation, the following conclusions were presented by the experimenters at the farm:

1. Juice from wheatgrass grown with Healthy Plant water had a superior taste; it was sweet and rich tasting, a pleasure to drink.
2. Healthy Plant juice had a lower sugar content than the control's juice, as measured by the Brix test. (This result is actually surprising, since the experimental wheatgrass tastes sweeter.)
3. Wheatgrass growing with Healthy Plant water showed a 10 percent increase in growth rate (when compared to the control).
4. Healthy Plant wheatgrass has a stronger leaf and root structure and is heartier and more vibrant (see fig. C.6 and fig. C.7 below).
5. The wheatgrass from treated flats produced an average of 60 percent more juice (in comparison with the control).

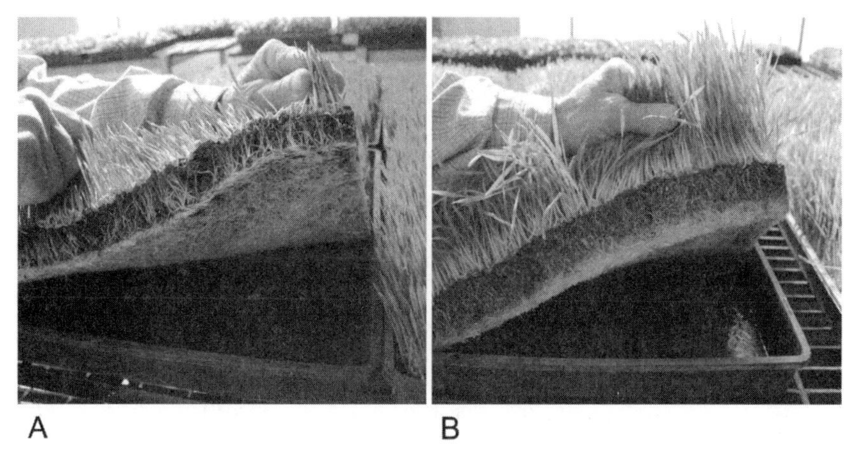

Fig. C.6. The roots of wheatgrass grown in a control group (A) are compared in these images to those grown with water infused with Healthy Plant energy pattern (B).

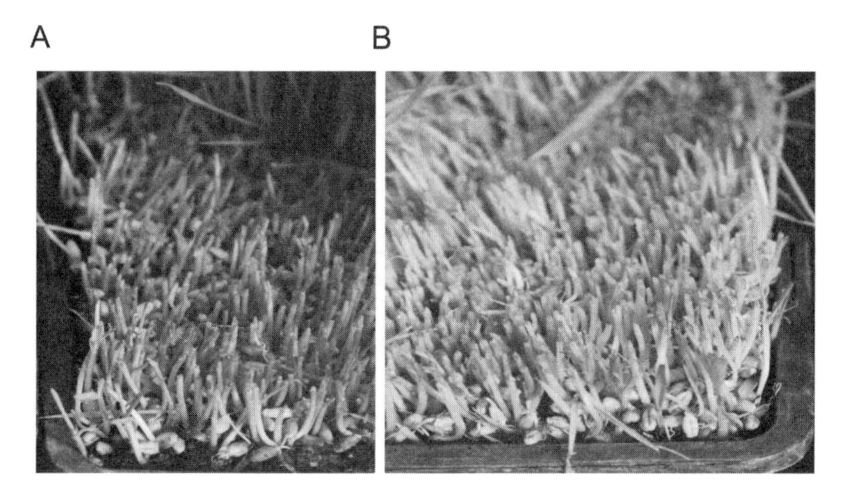

Fig. C.7. Wheatgrass crops in control group (A) are compared to those grown with water infused with Healthy Plant energy pattern (B).

6. Wheatgrass grown with Healthy Plant water had a longer shelf life, while maintaining good flavor. (Normally, wheatgrass turns bitter after a few hours. Healthy Plant wheatgrass maintained its good flavor for up to two days.)

Experiments with Onion and Rice Seeds at Oregon State University

In July 2015, experiments were conducted at the Seed Laboratory at Oregon State University in Corvallis, Oregon. Experimental seeds were divided into five groups and watered with infused water diluted in different proportions, with the following set of concentrations of infused water in the final watering mixture, given in percentages of 100, 10, 1, 0.1, and 0.01. The best results for the seeds' germination, both for rice and onion seeds, were observed with a dilution of 0.1 percent. The findings are presented below (see tables C.2 and C.3).

This experiment with seeds was done according to standard agricultural protocol. The seeds were planted on July 1, 2015, and kept in a controlled temperature and humidity. The first count was performed on July 8 and the final count was done on July 16, 2015.

In Table C.2, the results of the old onion seeds' germination are presented. Watering the seeds using water infused with the VFT formula Healthy Plant produced an increased germination rate of up to 164 percent.

TABLE C.2

	Initial amount	First count	Final count	Germinated seeds, %	Compared to control, %
Control water	99	22	28	28.3	100.0
Infused water	95	32	44	46.3	164.0

Table C.2. This table gives the results of the Healthy Plant formula experiment, which evaluated the germination rate of old onion seeds. The seeds were planted July 1, 2015, and the first count was conducted on July 8, 2015; the final count was done on July 16, 2015.

Table C.3 presents the results of the old rice seeds' germination. Watering seeds using water infused with the ETI formula Healthy Plant produced an increased germination rate up to 193 percent.

TABLE C.3

	Initial amount	First count	Final count	Germinated seeds, %	Compared to control, %
Control water	99	9	14	14.1	100.0
Infused water	99	21	27	27.3	192.9

Table C.3. Results of the Healthy Plant formula experiment on the germination of old rice seeds are given in this table. The seeds were planted July 1, 2015; the first count was conducted on July 8, 2015; the final count was done on July 16, 2015.

Experiments with Infused Fertilizer at Angel's Organic Farm

In 2018, Angel's Organic Farm conducted a pilot study with a fertilizer—Pacific Gro Oceanic Hydrolysate—that was infused and added to growing hemp plants. Two different types of hemp, Special Sauce and Sour Space Candy, were planted in four rows, with twenty-five plants in a row, each spaced four feet apart, and five feet between each row. The plants were given the same amount of fertilizer once a week for the entire growing season and only water for the last three weeks. The fertilizer was infused with the formula CSBMPCP, which combines the energetic imprints of calcium, silicon, boron, magnesium, phosphorous, cobalt, and potassium. In three months, the plants were measured to record the stem diameter, height, and width of each plant. Once the plant reached maturity, the base was cut and the entire plant was dried in a climate-controlled room. The individual plants were stripped of the stem to obtain the total extractable biomass weight.

Fresh leaf samples from the top (new growth) and lower portion (older leaves) of each plant were collected and analyzed. The samples were processed to obtain sap, from which a leaf extract analysis was performed. The results of the analysis are presented in figure C.8 on

Results of the hemp growing with the infused fertilizer		
Plant	Space Candy	Special Sauce
	% over control	
Brix	14	7
Manganese	17	7
Zinc	29	17
CBD	7	14
Average Weight	0	84

page 162.

Fig. C.8. Special Sauce hemp plants that were grown with treated (left) and untreated (right) fertilizer are compared in this set of photos.

The Vital Force Technology formula showed high efficacy for Special Sauce hemp, yielding 84 percent more biomass compared to the control. The research also demonstrated an increase in total CBD yield, at the level of 14 percent more per unit of weight, compared to the control.

APPENDIX D

Phantom Atom Effects on Animals and Human Cells

THE RESULTS of the HRV tests of Jeffrey Marrongelle support the notion of subtle energy as a regulator of all physiological processes in the body.[1] Table D.1 presents some examples of his case studies.

TABLE D.1

Paul D. 7.09.01, Male				
Condition–MS				
Post-test—15 minutes after dose of VitalForce formula				
	Pre-test	Post-test	Difference	
Parasympathetic	–3.0	–0.5	+2.5	VS*
Sympathetic	1.0	0.0	–1	S†
Heart Rate—supine	60	58	–2	
Heart Rate—upright	87	77	–10	VS
Tension Index—supine	156	70	–86	VS
Tension Index—upright	343	179	–164	VS
Optimum Variability (POV)—supine	7	16	+9	VS
Optimum Variability (POV)—upright	3	9	+6	S

*VS—Very Significant
†S—Significant

Pat M. 6.27.01, Female				
Condition—Sleep Apnea, Gallbladder				
Post-test—30 minutes after dose of VitalForce formula				
	Pre-test	Post-test	Difference	
Parasympathetic	–2.5	–0.0	+2.5	VS
Sympathetic	0.5	–0.5	–1	S
Heart Rate—supine	69	58	–11	S
Heart Rate—upright	72	64	–8	S
Tension Index—supine	275	47	–232	VS
Tension Index—upright	788	108	–680	VS
Optimum Variability (POV)—supine	11	27	+16	VS
Optimum Variability (POV)—upright	1	25	+24	S

Table D.1. Shown in these tables are examples of HRV test results after taking trace minerals infused with Stress Relief subtle energy pattern.

We see that in both patients almost all parameters measured by the HRV test were significantly improved after taking a water solution of trace minerals infused with the subtle energy formula Stress Relief.

The next diagram (fig. D.1) shows other examples of changes in parameters of the SNS and PSNS of patients in upright and supine position, as measured by the HRV test.

Subtle Energy Effects on Animals

During the experiments at the Laboratory of Behavioral Pharmacology at the University of Latvia in Riga, all mice were drinking water with trace minerals, using a concentration of 2 drops per 250 milliliters of water. Mice drank 4 to 5 milliliters of water per day, which corresponds to approximately 66 drops per day of trace minerals for a human weighing 160 pounds.

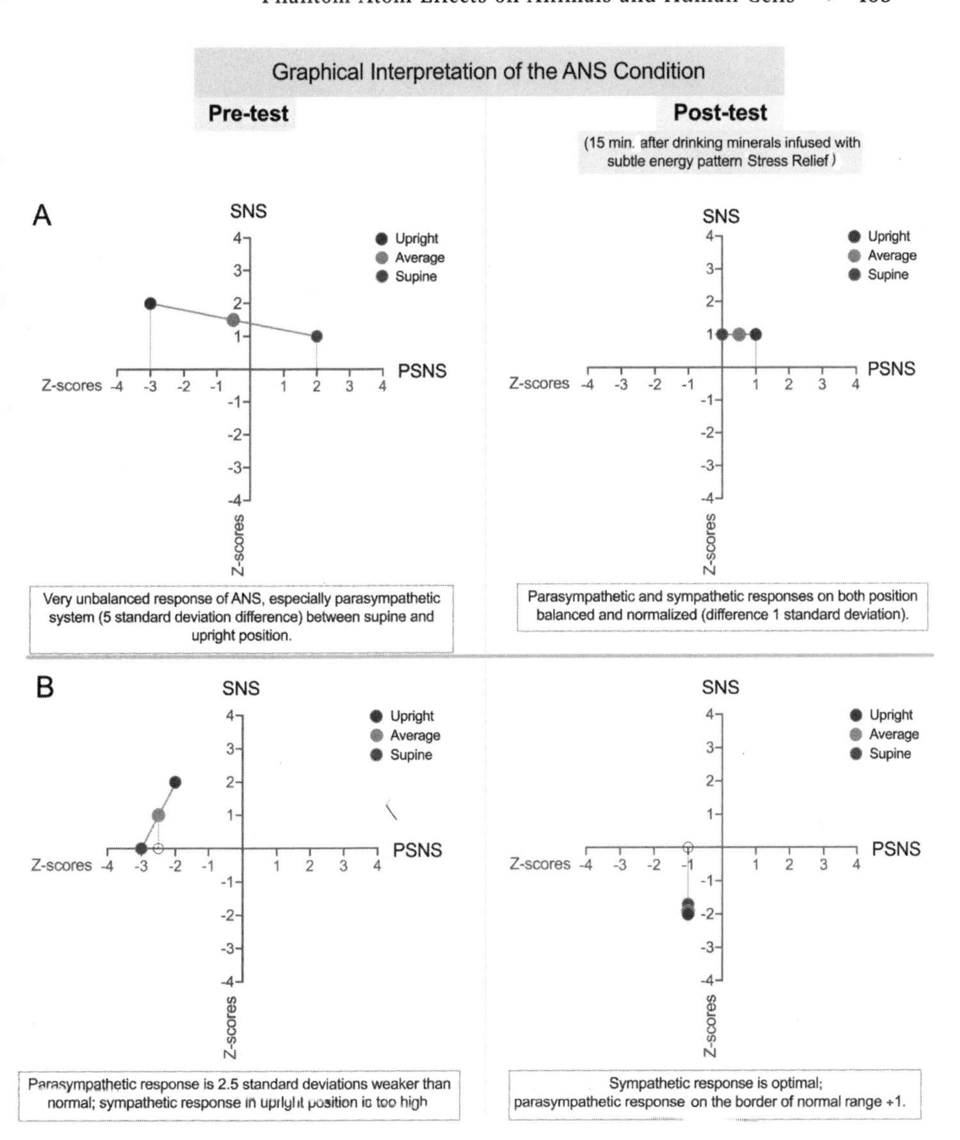

Fig. D.1. Two examples are given here (labeled patient A and patient B) of results from an HRV test. These results demonstrate the effectiveness and speed of action of the Stress Relief subtle energy pattern. (Note, z-score represents standard deviation.)

In three experiments performed during nearly three years, mice were divided into four groups according to the setup illustrated in figure D.2 on page 166.

Open Field Test

40 mice in each experiment were divided into 4 groups:

- **2 control groups** drank plain water with un-infused trace minerals and
- **2 experimental groups** drank water with trace minerals infused with Stress Relief

2 Control Groups:

20 mice that **WERE NOT** drinking Stress Relief. Divided into **2 groups of 10**:

- **10 mice** that **didn't** go through **stress** test **(Group #1)**
- **10 mice** were **stressed** using forced swimming stress (FSS), or predator odor stress (POS), or applying a combination of both stress factors **(Group #3)**

2 Experimental Groups:

20 mice that **WERE** drinking Stress Relief. Divided into **2 groups of 10**:

- **10 mice** that **didn't** go through stress test **(Group #2)**
- **10 mice** were **stressed** using forced swimming stress (FSS), or predator odor stress (POS), or applying a combination of both stress factors **(Group #4)**

Fig. D.2. These two columns describe the setup of an experiment that evaluated two control groups and two test groups in an Open Field Test.

Two control groups, with 10 mice in each, did not drink Stress Relief. One of these two control groups also did not go through the stress test (this group is labeled Group 1). The mice of the second control group were stressed using a forced swimming stress (FSS) test, described in figure D.3, which includes five minutes of swimming in a high-sided container of water and a predator odor stress (POS) test (this group is labeled Group 3).

Fig. D.3. This is a forced swimming stress (FSS) test, conducted as part of an experiment to evaluate the effectiveness of subtle energy patterns on laboratory mice.

Two experimental groups, also with 10 mice in each, were drinking water with trace minerals infused with Stress Relief. The mice of one experimental group did not go through the stress test (that is Group 2), while the mice of the second experimental group were stressed using FSS and POS tests (Group 4).

The software program that was used, Spontaneous Motor Activity Recording and Tracking (SMART), allowed automatic tracking of the position and movements of the mice and calculations of:

- The distance walked
- Central zone (no fear zone) crossings (shown in fig. D.4)
- Time spent in the central zone (no fear zone) (fig. D.4)
- Corner zone crossing (fig. D.4).

Open Field

The **SMART** (**S**pontaneous **M**otor **A**ctivity **R**ecording and **T**racking) software allows automatic tracking of the position and movements of the mice and calculations of:
- The distance walked
- Number of zone crossings
- Number of Central Zone (Fear Zone) crossings
- Time spent in the Central (Fear) Zone

Fig. D.4. The layout of the Open Field Test included several zones, or testing areas, and this is shown alongside a list of the main variables registered by SMART software.

Figure D.5 on page 168 presents an example of the mice's tracks recorded by SMART software in the Open Field Test.

The experiment was divided into three parts. In the first part mice were tested for nine days using only FSS test mode. In the second part, POS mode was applied. In this stage, three experiments with different

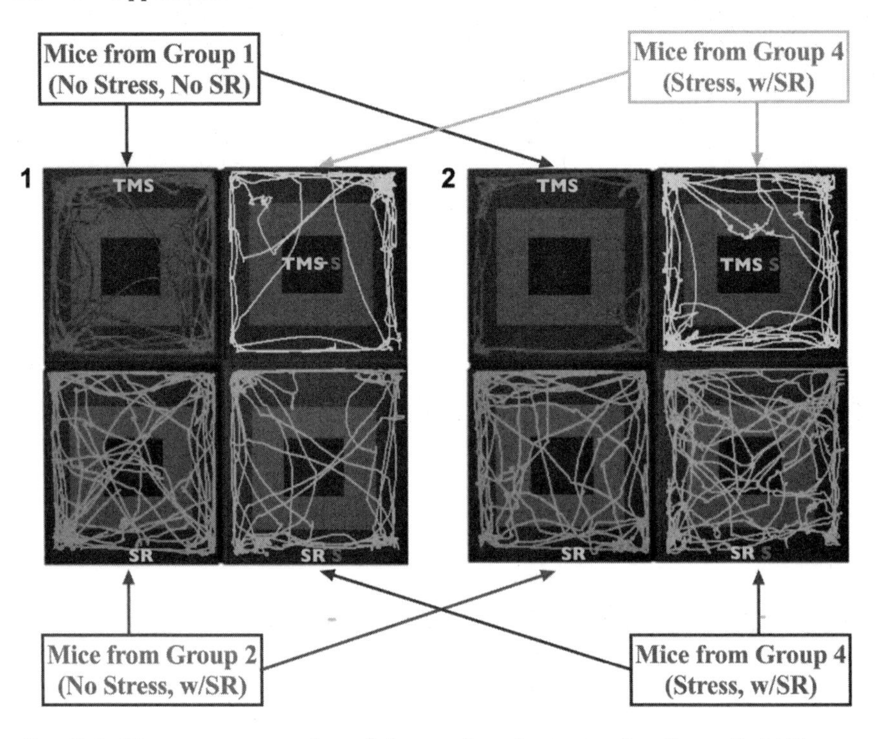

Fig. D.5. These are examples of the tracks of mice in the Open Field Test, as recorded by SMART software.

time schedules (7, 14, and 28 days) were carried out to test for time-effect relationships. In the last stage of this test, one thirty-two-day experiment was done with applying both stress factors, FSS and POS.

The results of the most important data measured are presented in the charts below. The first chart (fig. D.6) illustrates the distance walked by the non-stressed and stressed mice of the control groups that did not drink Stress Relief. We can see that stress definitely diminished these mice's activity. One can see that mice from both groups—those who were drinking Stress Relief (the non-stressed experimental Group 2 and the stressed experimental Group 4)—were much more active than the mice of both the non-stressed and stressed control groups. When we compare the number of central zone crossings of mice drinking Stress Relief with the same parameter of the control group (fig. D.7), we can see, again, that mice of both groups, stressed and non-stressed, performed noticeably more crossings than mice of the control groups.

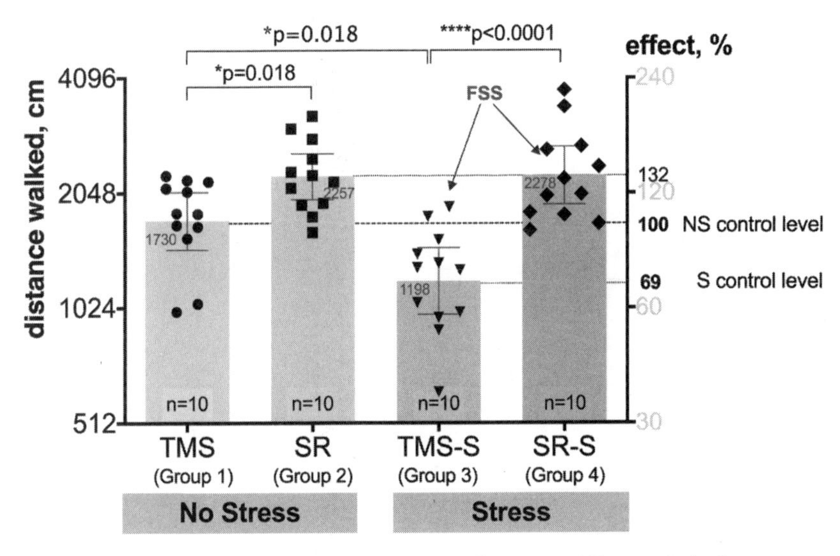

Fig. D.6. This bar chart compares the influence of Stress Relief energy pattern on distance walked by non-stressed and stressed mice in an Open Field Test. In this chart, FSS refers to forced swim stress, TMS is trace minerals solution, NS is non-stress, and S is stress.

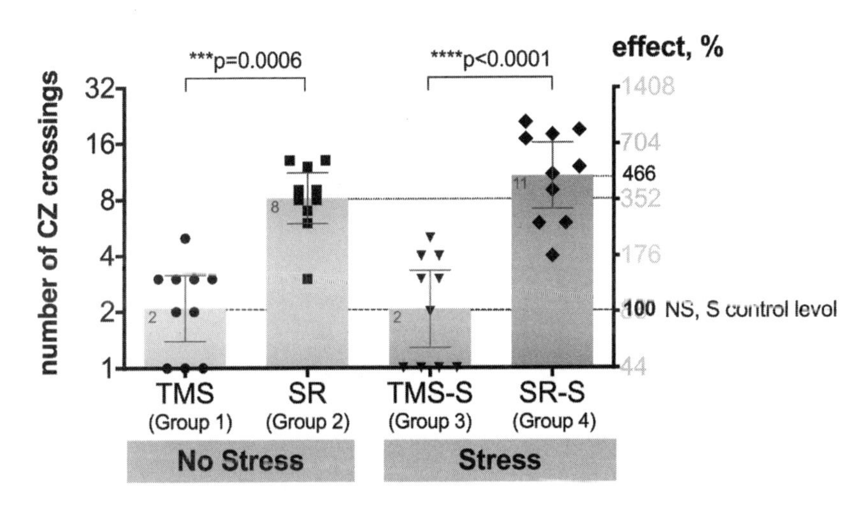

Fig. D.7. Influence of Stress Relief energy pattern on central zone (CZ) crossing by non-stressed and stressed mice in an Open Field Test. In this chart, TMS refers to trace minerals solution, NS is non-stress, and S is stress.

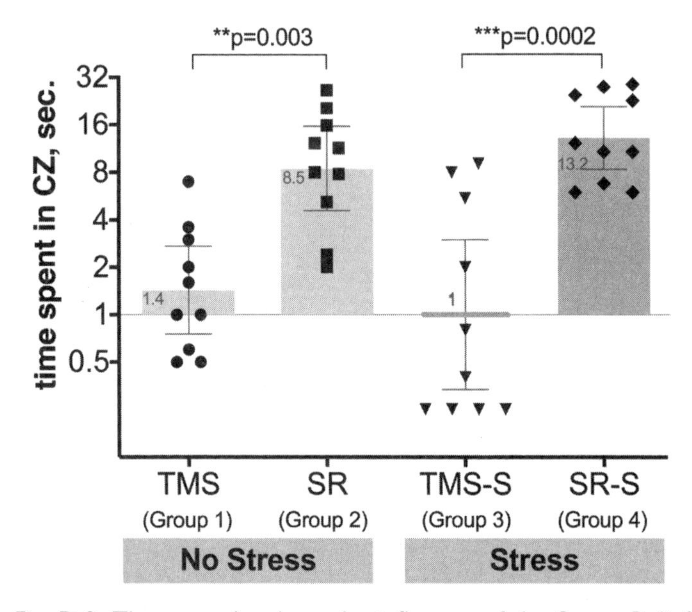

Fig. D.8. These results show the influence of the Stress Relief energy pattern on the amount of time spent in the central zone (CZ) by non-stressed and stressed mice in an Open Field Test. In the chart, TMS refers to trace minerals solution, NS is non-stress, and S is stress.

Mice of both groups drinking water with the Stress Relief formula also significantly outperformed non-stressed mice of the control group in time spent in the central zone, which is an indicator of their less stressed behavior (fig. D.8). Mice of the stressed group, represented as the SR-S group in figures D.6 and D.7, spent sufficiently longer time in the central zone (between 6 and 29 seconds) than the stress control group (from 0.5 to 9.1 seconds).

The parameters presenting the mice's fear level show an especially large difference between the behavior of mice drinking Stress Relief and those not drinking water with Stress Relief. The mice drinking Stress Relief made significantly more fear-zone crossings (fig. D.7) and spent substantially more time in that zone (fig. D.8). The stressed mice drinking Stress Relief dramatically outperformed all groups, including non-stressed mice drinking Stress Relief—in fact, the difference with stressed mice not drinking Stress Relief exceeded 300 percent!

Summarizing these results, we may conclude that this experimental study proved the Stress Relief formula stimulates a strong anti-stress response. Since exposure to the Stress Relief subtle energy pattern not only reverses stress-induced hypomotility but also increases the number of fear-zone crossings, it can be concluded that this energetic formula reduces the fear and stress of the treated animals.

Along with the Open Field Test, the blood glucose level in all mice groups was measured to determine physiological changes produced by stress (fig. D.9).

Tests showed that mice in the stressed group who were not drinking water with Stress Relief had significantly elevated levels of glucose in the blood—30 percent higher than in the control group. The stressed mice that drank water with Stress Relief had the same glucose level as the non-stressed mice in the control group. Test results showing normalization of the blood glucose level of the stressed mice group to

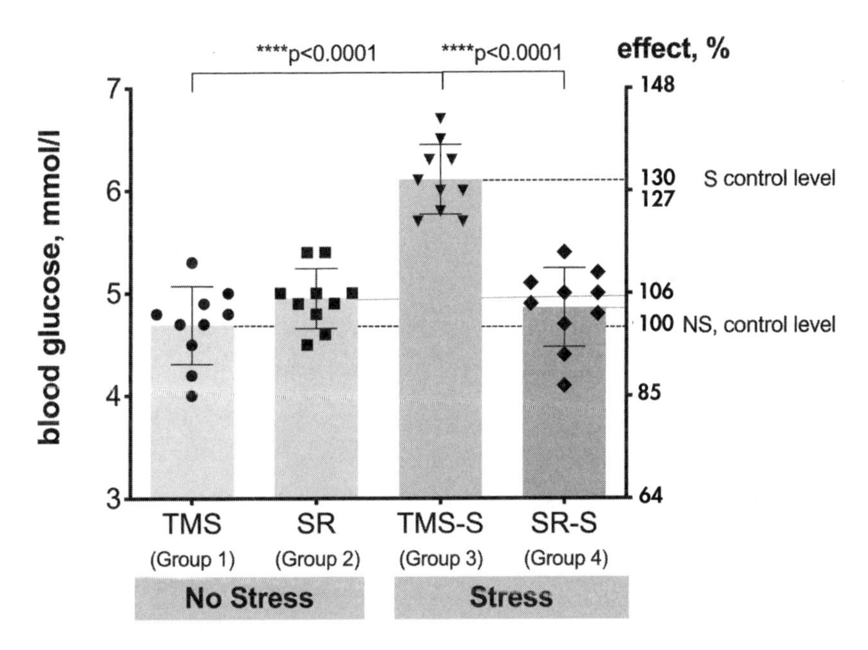

Fig. D.9. These results show the stress-induced rise of blood glucose in mice and its normalization after the use of Stress Relief energy pattern. In this chart, TMS is trace minerals solution, NS is non-stress, and S is stress.

the level of the non-stressed control group also support the efficacy of the Stress Relief formula.

It is evident that Stress Relief formula possesses some ergo-tropic features that serve as the basis of fight-or-flight response to acute stress, mobilizing energetic resources to active behavior. A similar, even more pronounced, effect was observed in animals that were exposed to short-term stress (five minutes of forced swimming, ten minutes exposure to predator odors, or a combination of both stress factors). Stress Relief not only reversed stress-induced hypomotility and activated locomotor functions of animals but also increased central zone crossing activity, time spent in the central zone, and the number of visits to the "stress corner," where the odor stress factor was applied, indicating stress-suppressive action (figs. D.5–8). The presented set of observations indicates that the Stress Relief subtle energy pattern has a fear- and stress-reducing effect in treated animals under different stressful situations, which is also clearly and strongly supported by results showing complete normalization of stress-induced elevation of blood glucose level (fig. D.9).[2]

Experiments on Human Dermal Fibroblast Cells

In the research conducted at Beech Tree Labs in Rhode Island and at the Department of Pharmacology at the University of Latvia, cells were seeded at a concentration of 50,000 cells per well, in 2 milliliters of medium in 6-well plates. Each medium infused with one of the energy patterns was placed in its own 6-well plate. Growth was measured by exposing the cells to a chemical called Alamar Blue. As the cells grew, Alamar Blue was metabolized and a florescent tag was formed. This tag was then measured by a plate reader to see the relative growth rate of the sample, as compared to the control. To create stricter conditions, and thereby display more prominently the potential effect of subtle energy patterns, all of the HDF cells were put in a condition of food deprivation. To generate this condition, the growth medium was not changed throughout the entire course of the experiment, so that the cells grew until they began to die.

Figure D.10 presents a comparison of the growth rates of cells during the last stage of growth (Day 6) and after they began to die (Day 9). One can see that the growth rate of cells grown in all of the energy-infused media was significantly higher than the control during the growth stage. The best result (82 percent more than in the control) was demonstrated by the energy pattern called Cell Longevity, which was created with VFT specifically for the support of the cells' functioning. The Peak Performance energy pattern, known for enhancement of physical endurance, also significantly stimulated the cells' growth and viability.

Please note that cells in the energy-infused media died much more slowly than the control cells: on Day 9, compared to the control, there were 83 percent more live cells in the medium infused with Stress Relief and 50 percent more in the media infused with Cell Longevity and Peak Performance. This preserving effect of the subtle energy patterns on the

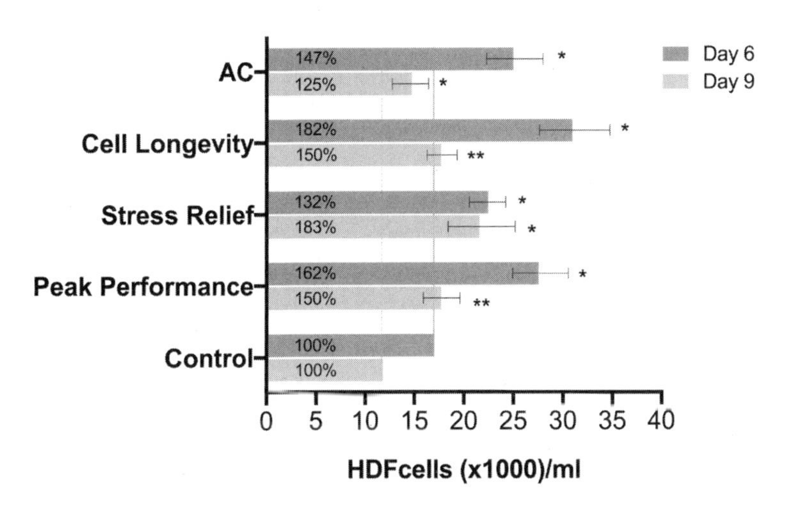

Fig. D.10. Here is shown the influence of four different energy patterns on HDF cell growth, represented by percentage of growth over the control group, as measured on Days 6 and 9. The gray dotted lines represent control levels. The asterisks adjacent to the bars in the chart represent the p-values, which are the result of hypothesis tests that show the statistical significance of these results; here, * = $p < 0.05$ and ** = $p < 0.01$, each compared to their corresponding control group.

cells situated in the toxic environment was a surprise for the researchers. The most significant effect was produced by the Stress Relief formula that was targeted to influence the body on the systemic level.

Linda Klimavičiusa, Ph.D., at the Department of Pharmacology at the University of Latvia tested subtle energy's ability to influence a cell's viability and the mitochondrial membrane potential under food deprivation conditions.[3] In addition, the researcher studied subtle energy's effects on cells when potent mitochondrial toxins were present (see table D.2).

TABLE D.2

Control cells		Experimental cells
Day 1	Growth medium + 10% FBS	Growth medium + 10% FBS
Day 2	Growth medium + 10% FBS	
Day 3 (Food deprivation)	As on Day 2, no FBS	As on Day 2, no FBS
Day 6	Test using MTT colorimetric assay and Tecan spectrometer	

Table D.2. This table describes the setup of an experiment that evaluated cell viability and mitochondrial membrane potential under food deprivation conditions, plus the presence of strong mitochondrial toxins.

The humn embryonic kidney cell line (HEK-293) was used for this experiment. On the first day, cells were grown in 96 well plates (with 100 microliters of medium per well) using a medium (DMEM/GlutaMAX-1) containing 10 percent fetal bovine serum (FBS). On the second day, the medium and serum in the experimental plates were changed for the plates with the medium and serum that were infused with Cell Longevity or Peak Performance. In the control, regular medium and serum were used. On the third day, the media were changed again to the same media without FBS to create serum deprivation conditions that lasted for the next three days before testing. Then, using the Tecan spectrometer, cell viability was tested by the MTT colorimetric assay, which uses the MTT tetrazole compound to determine the cellular respiration and metabolic activity of a sample. The cell viability of the control cultures was taken as 100 percent.

Results of a comparison of the cells grown in energy-infused media and the control are presented in figure D.11.

The Cell Longevity energy pattern, like in the previous experiment, produced the best effect on cell viability, at 46 percent higher than in the control, and Peak Performance showed a 21 percent increase in comparison with the control.

The next experiment was done with the addition of small amounts of mitochondrial toxins (besides previously described serum deprivation conditions). Two well-known types of toxins were used: MPP+ (1-methyl-phenylpyridine, 1.5 mM) and Rotenone (1 μM). See figures D.12 and D.13 on page 176.

In the case of the Rotenone toxin (fig. D.13), Cell Longevity

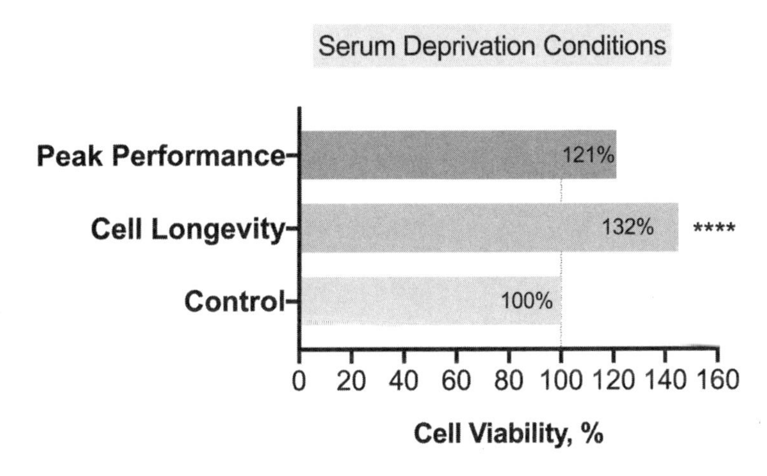

Fig. D.11. This chart shows the influence of Cell Longevity and Peak Performance energy patterns on cell viability. The cells were in medium containing each energy pattern for three days. The gray dotted line represents the control level. and the p-value, which is the result of a statistical test, is reported using asterisks, where **** = $p < 0.01$, compared to the control group.

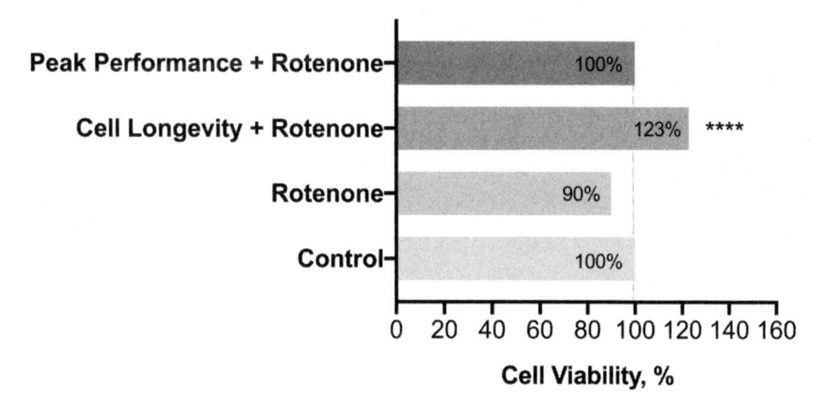

Fig. D.12. This chart shows the influence of Cell Longevity and Peak Performance energy patterns with Rotenone toxin on cell viability. The cells were in medium containing each energy pattern for three days. The gray dotted line represents the control level, and the asterisks refer to a statistical analysis, where **** = $p < 0.01$, compared to the group treated with Rotenone alone.

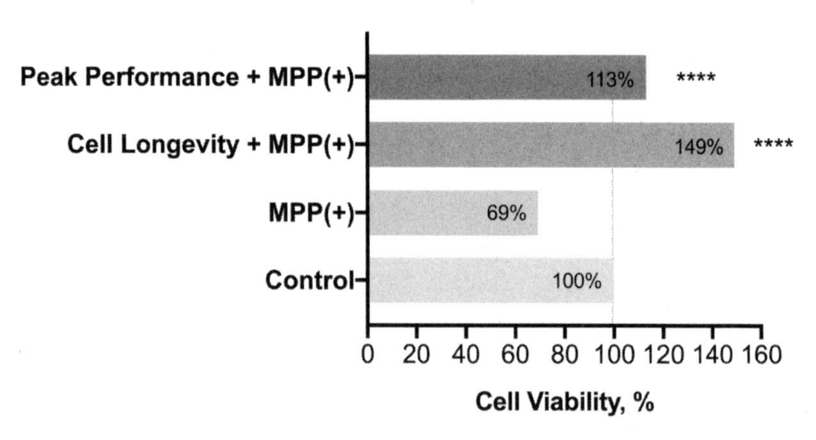

Fig. D.13. Influence of Cell Longevity and Peak Performance energy patterns with MPP toxin on cell viability. The cells were in medium containing each energy pattern for three days. The gray dotted line represents the control level, and the asterisks refer to a statistical analysis, where **** = $p < 0.01$, compared to the group treated with MPP alone.

improved the viability of cells more than in the non-poisoned control, while the Peak Performance energy pattern enhanced the cells' viability to equal that of the control cells. In the infused media, the viability of cells was better (for both energy patterns)—not only in comparison with the control containing mitochondrial poison but also in comparison with the non-poisoned control.

Figure D.14 shows a result of the measurement of mitochondrial membrane potential in media containing MPP+ poison and infused with Cell Longevity or Peak Performance, in comparison with non-infused controls. The measurement shows that cells in subtle energy–infused media had higher membrane potential than even the non-poisoned control with regular growth media.

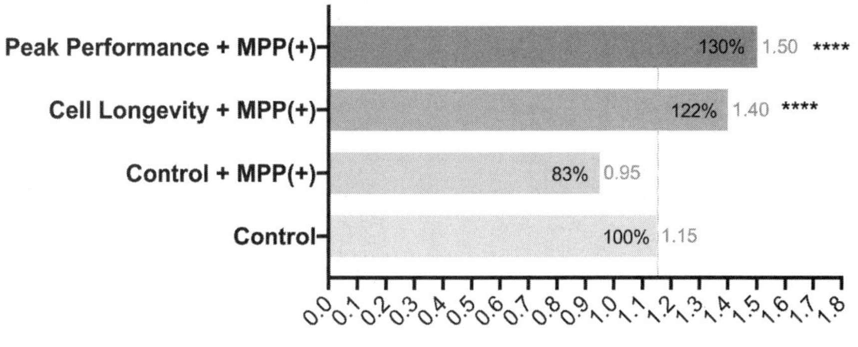

Fig. D.14. This chart shows the influence of Cell Longevity and Peak Performance energy patterns with mitochondrial toxin MPP(+) on mitochondrial membrane potential. The cells were in a medium containing each energy pattern for three days. The gray dotted line represents the control level, and the asterisks represent the results of a statistical analysis, where **** = $p < 0.01$, compared to the control + MPP(+) group.

Experiments on Genes

At Beech Tree Labs in Providence, Rhode Island, experiments were conducted in vitro using HDF cells. The experimental cells were grown for twenty-four hours in a medium infused with a specific subtle energy pattern created with the help of VFT. Then, using a sophisticated technique utilizing microarrays tailored to a specific disease or a cell process (like inflammation, immune response, etc.), the activity of the cell's genes was measured.

I need to mention that growing cells for twenty-four hours in an energy-infused media is not a sufficient period of time to significantly influence the cell's physiology. In this first experiment, the researchers were not aware of that. As we will see later, growing cells in the energy-infused media for a longer time produced much more pronounced effects. Nevertheless, even in this twenty-four-hour experiment, statistically significant evidence of subtle energy's effect on the gene regulation was observed.

Table D.3 presents the results of the experiment in which HDF cells were exposed to a subtle energy pattern aimed at reducing pain.

Several genes responsible for the reaction of the body to pain were down-regulated. One of these genes—the very important gene called tachykinin receptor 1 (TACR1)—was down-regulated significantly (up and down regulation for more than twofold is considered significant). TACR1 is the gene that encodes for the receptor neurokinin 1. When stimulated by being bound with its activator, TACR1 is associated with the transmission of stress signals and pain as well as the contraction of smooth muscles and inflammation. Down-regulating this gene would lead to less receptors being produced and would potentially result in a diminished response to the stimulating pain, decreased inflammation, and less stress on the body. Receptor neurokinin 1 can be found in both the central and peripheral nervous systems.

The next experiment conducted at Beech Tree Labs answered an important question: Can subtle energy enhance drug effectiveness? The experiment was done on a drug called M2, using subtle energy pattern

Anti-Aging, which intends to slow down the aging of the human body's cells. M2 is manufactured by Eudaemonic under the name RVI (rubeola virus immunomodulator). It is an inactivated rubeola virus combined with histamine phosphate in physiological saline. M2 aids in the healing process leading to the decrease or elimination of myofascial inflammation.

TABLE D.3

Gene Description	Gene Name	Fold Change	P-value
Bradykinin receptor B1	BDKRB1	–1.10	0.003
CD4 molecule	CD4	–1.85	0.004
5-hydroxytryptamine (serotonin) receptor 2A	HTR2A	–1.47	0.011
Kv channel interacting protein 3, calsenilin	KCNIP3	–1.29	0.029
Mitogen-activated protein kinase 1	MARK1	–1.14	0.002
Mitogen-activated protein kinase 14	MARK14	–1.24	0.038
Mitogen-activated protein kinase 3	MARK3	–1.16	0.017
Purinergic receptor P2X, ligand-gated ion channel 4	P2RX4	–1.26	0.000
Prostaglandin E receptor 1 (subtype EP1), 42kDa	PTGER1	–1.38	0.026
Prostaglandin E synthase	PTGES	–1.08	0.004
Sodium channel, voltage0gated, type IX, alpha subunit	SCN9A	1.16	0.004
Tachykinin receptor 1	TACR1	–2.14	0.036
Toll-like receptor 4	TLR4	–1.15	0.002

Table D.3. The results in this table show what happened to HDF cells exposed to subtle energy Pain formula for twenty-four hours.

In this experiment, M2 was added to the growth medium of cells. Another group of cells grew in the same medium but was infused with subtle energy pattern Anti-Aging. This time cells were grown in the energized medium for seventy-two hours. To explore any possible effect of subtle energy on the drug, M2 was added to the cells grown in the energized medium. Then, the genetic activities of the cells of both groups were compared (table D.4).

The Anti-Aging formula enhanced M2's effect on gene C3 by 58 percent and the effect on CX3CL1 by 136 percent. Additionally, subtle energy significantly (more than threefold) up-regulated genes that M2 alone does not. We can see that growing cells in the energized medium for a long time leads to a significantly stronger effect on the gene expression. Interestingly, most genes being regulated with M2 (and M2 with subtle energy) are genes involved in the immune system.

M2 has a significant effect on gene regulation.
M2 gene regulation was noticeably enhanced with the addition of subtle energy pattern Anti-Aging.
Subtle energy effectively regulates genes, where M2 does not.

It needs to be pointed out here that gene BUBIB, promoting a proper chromosome segregation, was down-regulated by M2 alone, which is not a good thing. An addition of the subtle energy pattern to M2 up-regulated this gene almost threefold back to normal.

An additional experiment conducted at Beech Tree Labs demonstrated that the subtle energy pattern copied from a drug (Adhesion-BTR, created at Beech Tree Labs) affects the same genes as the drug itself. The gene's activity of Adhesion-BTR was compared with activity of the subtle energy formula Adhesion-ETI, an energetic copy of the drug. Human aortic endothelial cells (AEC) were chosen for this research as they are sensitive to the adhesion. They were seeded at 125,000 cells and allowed to grow until their confluence. Once confluent (meaning a nice monolayer has formed), the cells were given 100 microliters of either PBS (control) or Adhesion formulation. The cells were grown for seventy-two hours, harvested, and RNA was collected. The genomics were run on a Human Extracellular Matrix and Adhesion Molecules array (Qiagen). Then, the genetic activities of the cells were compared.

TABLE D.4

		HDF cells exposed to M2 and Subtle Energy			
		M2 alone		M2 with Anti-Aging	
Gene Description	Gene Name	Fold Change	P-value	Fold Change	P-value
Budding uninhibited by benzimidazoles 1 homolog beta (yeast)	BUB1B	–4.47	0	–1.65	0.03
Complement component 1, s subcomponent	C1s			3.63	0
Complement component 3	C3	16.71	0.01	26.62	0
Caspase 1, apoptosis-related cysteine peptidase (interleukin 1b convertase)	CASP1	1.81	0.04	2.2	0
CD 163 molecule	CD163	1.8	0.01	2.11	0.02
Complement factor H	CFH	2.91	0.01	4.42	0
Collagen type III, a1	COL3a1	2.19	0	2.73	0
Chemokine (C-X3-C motif) ligand 1	CX3CL1	7.91	0.02	18.66	0
Chemokine (C-X3-C motif) ligand 16	CXCL16	3.03	0	4.31	0
Fc fragment of IgG, high affinity 1a, receptor (CD64)	FCGR1a	6.31	0.01	6.84	0
Myelin basic protein	MBP	–2.52	0.01	–2	0.01
Pannexin 1	PANX1	2.59	0	3.44	0
Ring finger protein 144B	RNF144B	4.57	0	4.97	0
Fc fragment of IgE, high affinity 1, receptor for, g polypeptide	FCER1g			3.12	0

Table D.4. The results presented here show what happened to HDF cells exposed to the subtle energy formula called Anti-Aging.

As we see in table D.5, protein-coding gene MMP9 (Matrix Metallopeptidase 9) is up-regulated with formula Adhesion-BTR 1.44-fold, to compare with almost twofold up-regulation with Adhesion-ETI formula. This is a positive response, as MMP9 is involved in the extracellular matrix and would help with adhesion. This experiment confirmed that the technologically enhanced energy pattern of a drug (Adhesion-ETI) affects the same genes as the drug itself (Adhesion-BTR) even more effectively.

TABLE D.5

	Adhesion-BTR		Adhesion-ETI	
Formula	Fold Change	P-value	Fold Change	P-value
MMP2	X	X	X	X
MMP9	1.4443	0.007	1.9734	0.0004
MMP11	X	X	X	X
MMP14	X	X	X	X

Table D.5. This table shows what happened to genes affected by the subtle energy formulas called Adhesion-BTR and Adhesion-ETI.

APPENDIX E

Pranic Healing and Electromagnetic Pollution

IN HIS EXPERIMENTS, Professor Joie Jones observed that the healing effects produced by the pranic healers were dramatically different when conducted in energetically clean versus energetically dirty environments. Jones performed his experiments in three distinctly different laboratories and found three very different results:

1. The conditioned lab: a laboratory that had been regularly cleaned by pranic healers (the "cleaning" referred to here is a special procedure these healers commonly used for energetic cleaning of the environment before each session).
2. The non-conditioned lab: a fairly new and well-maintained research environment.
3. The dirty lab: a laboratory where experiments on dissected animals and other similar experimentation had been conducted for years.

Here are the results Jones observed in the three laboratories (table E.1 on page 184):

TABLE E.1

	Conditioned lab	Non-conditioned lab	Dirty lab
Success Rate	88% (T = 854)	10% (T = 150)	0% (T = 150)

Table E.1. Experiments conducted in three distinctly different laboratories gave three very different results. Here, the success rate is the number of petri dishes with an increased survival rate, and the total number of experiments (T) is given in parentheses following the success rate.

In his first experiments with the VFT energetic patterns, Jones determined the energy pattern that provided the biggest increase of the survival rate of the damaged HeLa cells. We will refer to this as the healing (H) energy pattern. Similar to what was found in the pranic healing experiments, the maximum increase in the survival rate of cells in successful petri dishes occurred when the subtle energy–infused solution was added both before *and* after the radiation of the cells (see fig. E.1)

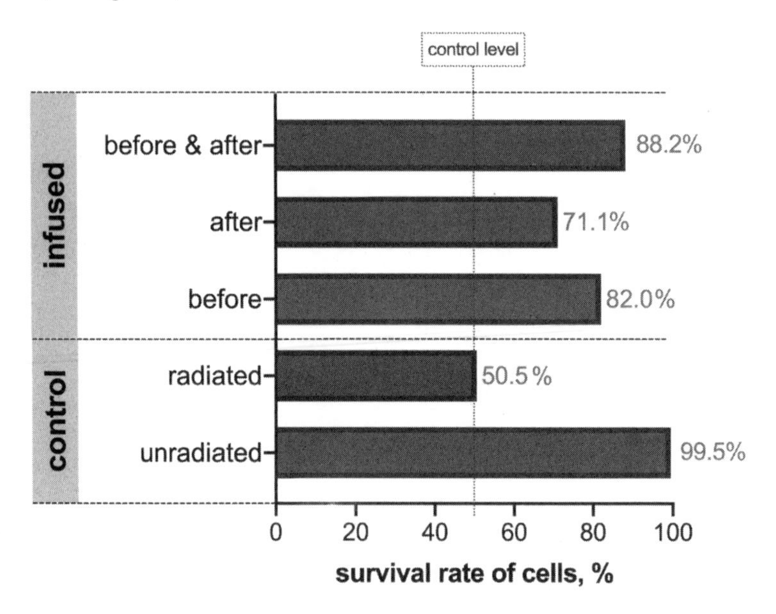

Fig. E.1. The results shown here compare the percentage of survival rate of cells that were infused with a healing subtle energy pattern in a conditioned lab.

Just as it was found in the pranic experiments of Jones, the success rate (the number of petri dishes with an increased survival rate) in this study was very different in the conditioned, non-conditioned, and dirty laboratories.

Energetically polluted environments clearly have a tremendous effect on the experiments. The results in table E.2 show that energetic pollution literally obstructs the capability of the subtle energy pattern–infused solution to increase the survival rate of cells subjected to gamma radiation. This fact was observed no matter how the healing energy was applied—by healers or through technological infusion. The fact that the subtle energy pattern generated by VFT equipment was much stronger than when projected by a pranic healer made no difference: it did not override the negative effects of the energetic pollution on the healing process.

TABLE E.2

	Pranic Healing	H Energy Pattern
Conditioned Lab	88%	89%
Non-Conditioned Lab	10%	11%
Dirty Lab	0%	0%

Table E.2. These results show the results of experiments that evaluated the survival rate of cells in different lab conditions. There were forty-five experiments conducted in each lab (all using the H energy pattern). The percentages refer to increased survival rate of cells. In the non-conditioned lab, cells show a slight increase in survival rate over the dirty lab. The survival rate in the conditioned lab, which was cleaned regularly by pranic healers, shows a significantly higher percentage of cell survival.

These results encouraged us to conduct additional experiments aimed at finding ways to protect the healing action of the subtle energy pattern infused into a Ringer's solution of several salts dissolved in water used for growing cells. In 2006, we set up an experiment aimed at finding a way of counteracting the negative effect of energetic pollution to create conditions where the healing power of subtle energy could be observed consistently and in any laboratory. An additional goal of our experiments was to shed some light on the energetic mechanisms of the human mind's power of intent.

The energy patterns intended to fight energetic pollution were created by pranic healers and subsequently recorded and produced by VFT. Each pranic healer concentrated his or her power of intent for ten to fifteen minutes on a bottle of concentrated trace mineral solution* in water. Energy patterns of this solution were copied using VFT, producing a new protective energy pattern (called P). The next forty-five experiments were conducted in each of the same three labs. This time, the Ringer's solution added to the experimental samples was infused with the P pattern, along with the healing subtle energy pattern H.

Table E.3 shows that the presence of the protective energy pattern dramatically increased the number of successful experiments in both the non-conditioned and dirty labs.

TABLE E.3

	Pranic Healing	H Energy Pattern	H & P Energy Patterns
Conditioned Lab	88%	89%	89%
Non-Conditioned Lab	10%	11%	87%
Dirty Lab	0%	0%	78%

Table E.3. This table shows results of the experiments of applying the H & P energy pattern to cells to counteract the energetic pollution of the labs. The first two columns show the results from the previous experiments shown also in table E.2. The right-most column shows the survival rate of the cells treated with the H & P energy patterns. The rates for the conditioned lab remain in the same range as the earlier experiments, but the non-conditioned and dirty labs show a significant increase in cell survival rate. Total number of experiments in each lab: 45 (all using H & P energy patterns).

*During the many years of experimentation with VFT, we discovered that trace mineral solutions are very effective carriers of the subtle energy patterns infused into them. The shelf life of these energy-infused minerals lasted for years.[1]

It is useful to note that adding the protective energy pattern made the average overall increase in the cells' survival rate in the dirty lab very significant statistically (table E.4).

TABLE E.4

	A / Non-Infused Solution	B / Non-Infused Solution	C / Infused Solution
Conditioned Lab	Control untreated	Control radiation treated	Treated before and after radiation
HP, ave all (n = 45)	99.7%	50.3%	82.9%
HP, ave suc (n = 40)	99.3%	50.4%	88.0%
HP, ave fail (n = 5)	99.5%	50.7%	50.7%
Non-Conditioned Lab			
HP, ave all (n = 45)	99.4%	50.1%	82.4%
HP, ave suc (n = 39)	99.5%	49.8%	88.1%
HP, ave fail (n = 6)	99.3%	50.3%	50.8%
Dirty Lab			
HP, ave all (n = 45)	99.2%	50.3%	79.5%
HP, ave suc (n = 35)	99.1%	50.5%	88.0%
HP, ave fail (n = 10)	99.3%	50.1%	51.0%

Table E.4. This table shows the survival rate of cells in three labs, where HP represents a healing energy pattern with a protective energy pattern.

After the success of creating a protective energy pattern, we attempted to create a cleansing pattern that could clean the entire space in the dirty lab. Utilizing the same protocol we used for the creation of the protective subtle energy pattern, we made a cleansing energetic pattern.*

In our first experiments with the cleansing energy pattern, we

*Later, energy patterns of several gemstones known for their cleansing abilities were added to this pattern.

recorded the pattern with the VFT equipment in a DVD audio format using the highest possible resolution. We played this DVD for twenty minutes through an audio system (no sound is heard, and only subtle energy of the cleansing pattern is emitted). The results are presented in table E.5:

TABLE E.5

	Pranic Healing	H Pattern
Conditioned Lab	88% (T = 854)	89% (T = 30)
Non-Conditioned Lab	10% (T = 150)	87% (T = 30)
Dirty Lab	0% (T = 150)	68% (T = 30)

Table E.5. This table shows the percentage of successful experiments after treatment of laboratory spaces with the cleansing energy pattern. The survival rate of cells is significantly increased for the non-conditioned and dirty lab conditions.

It should be noted that when Jones's lab was prepared for experiments for the first time, it took pranic healers several months to condition the lab, and an increase in the survival rate of cells was never observed in the dirty lab. In our experiment, after twenty minutes of energetic cleansing of the lab space, we observed an increase of successful experiments in the dirty lab from 0 percent to 68 percent.

To determine how long the energetic cleansing is effective, Jones repeated the experiments with radiated cells in all three labs at intervals of three days, from the time of the original cleansing. Figure E.2 presents the results.

In reviewing the data, one notices that the continued presence of energetic pollution in the dirty lab obstructs the healing energy much more than it does in the regular non-conditioned lab. More important, however, based on the experimentally derived data cited, we may conclude that the frontier science technology of programming subtle energy patterns is capable of sufficiently counteracting the negative effects of energetic pollution.

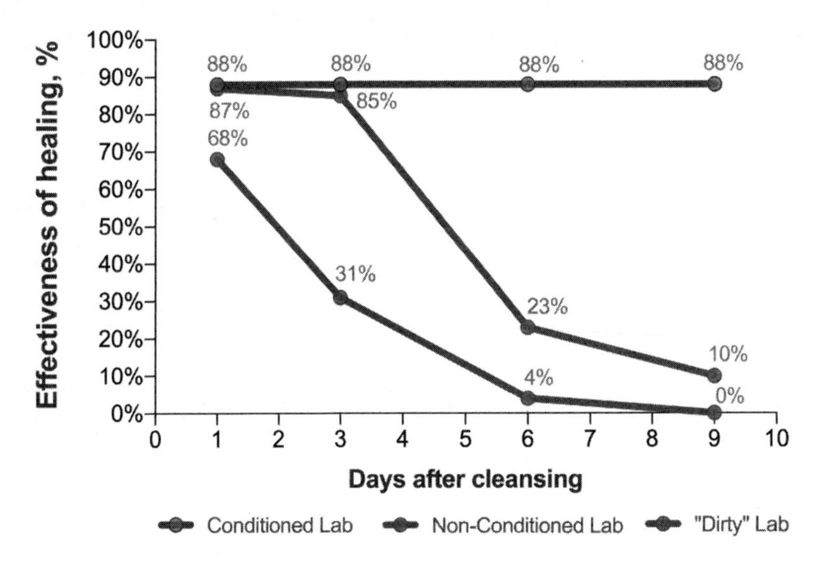

Fig. E.2. This line graph depicts the effectiveness and duration of energetic cleansing in different labs. The conditioned lab that was regularly treated with healing energy before the experiments shows the healing rate for cells maintains a static level. The healing rate fell off sharply in the non-conditioned lab as well as the dirty lab.

Electromagnetic Pollution: A Growing Threat to All Life on Earth

Jeffery Fannin, Ph.D., founder of the Center for Cognitive Enhancement in Glendale, Arizona, did a pilot research study to see how smartphones influenced electrical brain activity and whether specific subtle energy patterns might help the brain to maintain normal functioning in the presence of cell phone radiation. This pilot research was conducted with ten subjects using qEEG brain mapping equipment and software created by BrainMaster Technologies. The participants (fives males and five females, ranging from thirty-four to sixty-four years old) were normal, healthy individuals who were screened for absence of psychological and neurological abnormalities via qEEG brain mapping.

Let us look at the examples of the brain maps presenting the electrical brain activity when no cell phone was used (see fig. E.3 on page 190), after talking for three minutes on a smartphone positioned one to two inches from the head (fig. E.4a), and when the same subject was using the same phone with an attached piece of plastic infused with a subtle energy pattern created for normalization of electrical brain activity (fig. E.4b).

Colors on the brain maps present corresponding levels of electrical activity in the given area of the brain. The following color codes are used by the software of the qEEG system: green shows the normal amplitude of the brain's electrical activity, yellow/orange represents two standard deviations (SDs) above normal, red represents three or more SDs above normal, and light blue represents two SDs below normal.

Figure E.4a presents an example of the effect of a cell phone's

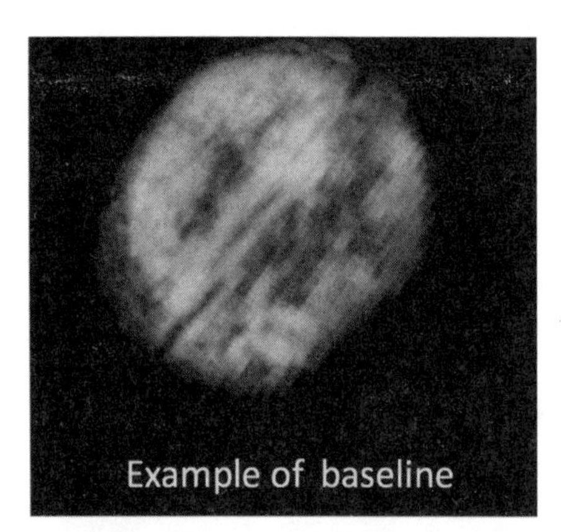

Fig. E.3. This image shows baseline brain activity, using qEEG brain mapping diagrams.

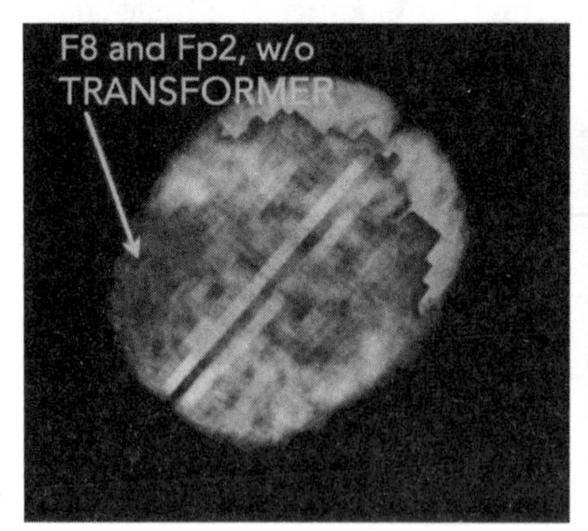

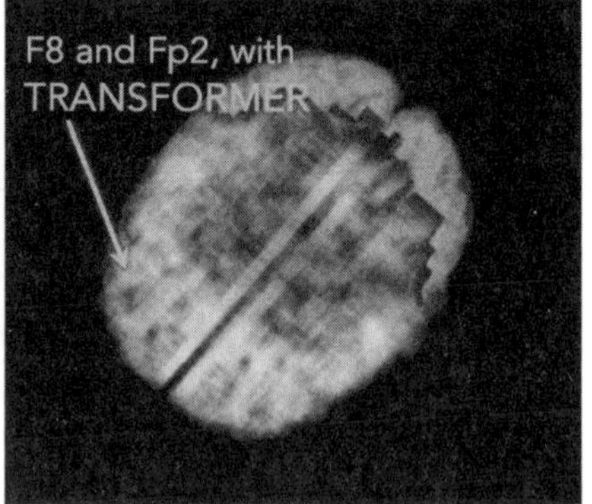

Fig. E.4a (left) and b (right). These images show the effect of a cell phone's radiation on the frontal lobe regions called F8 and Fp2, both with and without a transformer.

radiation on the right frontal lobe when the phone was used. The density of this dark spot shows the effects are in the range of three SDs above normal in the affected part of the frontal lobe. This area of the brain includes points F8 and Fp2, according to the qEEG system of classification. Excessive activity in this region of the brain has been shown to cause problems with working memory, such as spatial and visual problems, issues with gestalt (configuring objects and experience), trouble processing facial emotional expressions, and possible problems with sustained attention. The elevation of neuronal activity using the cell phone at Fp2 suggests a person may have less efficiency in both emotional attention and verbal expression.

As you can see from figure E.4b, the presence of the subtle energy pattern called Transformer normalizes the amplitudes of the brain waves in both points F8 and Fp2, as well as in all areas of the frontal lobe that were previously overstimulated (fig. E.4). The arrow in fig. E.4a (left) points to a scan of the frontal lobe regions of the brain called F8 and Fp2. This scan displays a dark spot that shows up when a cell phone is used without Transformer. The arrow in fig. E.4b shows that the dark spot is cleared when a cell phone is used with Transformer.

After analysis of all of the participants' brain maps, Fannin made the following conclusion: "Results of independent tests . . . demonstrated higher levels of activity in all frequency bands (delta, theta, alpha and beta) from the left temporal regions continuous to frontal locations when cell phone use was engaged WITHOUT the infused 'Transformer.' . . . More normal brainwave activity is present in the areas examined when using the cell phone WITH the infused Transformer."

References

Chapter 1. A Force We All Experience

1. Talbot, Michael. 1991. *The Holographic Universe: A Remarkable New Theory of Reality that Explains: The Paranormal Abilities of the Mind, The Latest Frontiers of Physics, and the Unsolved Riddles of the Brain and Body.* New York: HarperCollins.
2. Chin, Richard. 1995. *The Energy Within: The Science Behind Eastern Healing Techniques.* New York: Marlowe and Company.
3. Collinge, William. 1998. *Subtle Energy: Awakening to the Unseen Forces in Our Lives.* New York: Warner Books.
4. Eden, James. 1993. *Energetic Healing: The Merging of Ancient and Modern Medical Practices.* New York: Plenum Press.
5. Gerber, Richard M.D. 1988. *Vibrational Medicine: New Choices for Healing Ourselves.* Santa Fe: Bear & Company.
6. Lipton, Bruce. 2005. *The Biology of Belief: Unleashing the Power of Consciousness, Matter and Miracles.* Santa Rosa: Mountain of Love/ Elite Books.
7. Radin, Dean. 2006. *Entangled Minds: Extrasensory Experiences in a Quantum Reality.* New York: Paraview Pocket Books, a division of Simon & Schuster, Inc.
8. Radin, Dean. 1997. *The Conscious Universe: The Scientific Truth of Psychic Phenomena.* New York: HarperCollins.
9. McTaggart, Lynne. 2008. *The Field: The Quest for the Secret Force of the Universe.* New York: HarperCollins.
10. Jahn, Robert G., and Brenda J. Dunne. 1987. *Margins of Reality.* Orlando: Harcourt Brace Jovanovich.

11. Tiller, William A. 2007. *Psychoenergetic Science: A Second Copernican Scale Revolution*. Walnut Creek, Calif.: Pavior Publishing.
12. Laszlo, Ervin. 2004. *Science and the Akashic Field: An Integral Theory of Everything*. Rochester, Vt.: Inner Traditions.
13. Ni, Hua-Ching. 1979. *The Complete Works of Lao Tzu: Tao Teh Ching & Hua Hu Ching*. Santa Monica: Seven Star Communications Group.
14. Swanson, Claude. 2010. *Life Force, the Scientific Basis: Breakthrough Physics of Energy Medicine, Healing, Chi and Quantum Consciousness*. Tucson: Poseidia Press.
15. Meyl, Konstantine. 2011. *DNA and cell resonance: Communication of cells explained by field physics including magnetic scalar waves*. Erikaweg, Germany: Villingen-Schwenningen.
16. Eden, Donna. 2008. *Energy Medicine: Balancing Your Body's Energies for Optimal Health, Joy, and Vitality*. New York: Penguin Group.
17. Bruyere, Rosalyn L. 1989. *Wheels of Light: A Study of the Chakras*. Sierra Madre: Bon Productions.
18. Brennan, Barbara Ann. 1987. *Hands of Light: A Guide to Healing Through the Human Energy Field*. New York: Bantam Books.
19. Zuyin, Lu. 1997. *Scientific Qigong Exploration: The Wonders and Mysteries of Qi*. Malvern, Pa.: Amberleaf Press.
20. Connelly, Dianne M. 1971. *Traditional Acupuncture: The Law of the Five Elements*. New York: New Directions Publishing Group.

Chapter 2. The Nature of Subtle Energy

1. Trefil, James. 1993. "Dark Matter." *Smithsonian,* June.
2. Barbree, Jay, and Caidin Martin. 1995. *A Journey Through Time: Exploring the Universe with the Hubble Space Telescope*. London: Penguin Books.
3. Glanz, James. 2001. "First Direct Evidence of Negative Gravity." *National Post Online*. April 1. National Post website.
4. Cline, David B. 2003. "The Search for Dark Matter." *Scientific American,* March.
5. Urry, Meg. 2007. "The Secrets of Dark Energy." *Parade Magazine Sunday Supplement,* May 27.
6. Clark, Stuart. 2012. "No More Eureka Moments." *New Scientist,* January 9.

7. Kaplan, David E., Gordon Krnjaic, Keith Rehermann, and Christopher Wells. 2010. "Atomic Dark Matter." *Journal of Cosmology and Astroparticle Physics* vol. 2010.
8. Laszlo, Ervin. 2004. *Science and the Akashic Field: An Integral Theory of Everything*. Rochester, Vt.: Inner Traditions.
9. Powell, A. E. 1954. *The Astral Body*. London: Theosophical Publishing House.
10. Powell, A. E. 1956. *The Causal Body*. London: Theosophical Publishing House.
11. Powell, A. E. 1953. *The Etheric Double and Allied Phenomena*. London: Theosophical Publishing House.
12. Powell, A. E. 1956. *The Mental Body*. London: Theosophical Publishing House.
13. Dale, Cyndi. 2009. *The Subtle Body: An Encyclopedia of Your Energetic Anatomy*. Boulder: Sounds True.
14. Ni, Hua-Ching. 1979. *The Complete Works of Lao Tzu: Tao Teh Ching & Hua Hu Ching*. Santa Monica: Seven Star Communications Group.
15. Zuyin, Lu. 1997. *Scientific Qigong Exploration: The Wonders and Mysteries of Qi*. Malvern, Pa.: Amberleaf Press.
16. Xin, Yan, Feng Lu, Hongjiang, et al. 2002. "Certain Physical Manifestations and Effects of External Qi of Yan Xin." *Journal of Scientific Exploration* (Allen Press): 16.
17. Bowman, Carol L. 2013. "Energy Medicine—Science and Beyond." Life Vessel of the Rockies website.
18. Thie, John, and Matthew Thie. 2010. *Touch for Health*. Camarillo, Calif.: DeVorss & Company.
19. Harris, W. S., M. Gowda, J. W. Kolb, et al. 1999. "A Randomized, Controlled Trial of the Effect of Remote Intercessory Prayer on Outcomes in Patients Admitted to the Coronary Care Unit." *Archives of Internal Medicine,* 159 ed.
20. Jones, Joie P. 2006. "An Extensive Laboratory Study of Pranic Healing Using Contemporary Medical Imaging and Laboratory Methods." Invited presentation for the Seventh World Pranic Healers' Convention, May 12–14. Mumbai.
21. Jones, Joie P. 2006. "Pranic Healing: fMRI Measurements of Subtle Energy." A Think Tank Working Group Meeting on Biofield Energy Medicine, March 29–31. Bethesda.

22. Swanson, Claude. 2010. *Life Force, the Scientific Basis: Breakthrough Physics of Energy Medicine, Healing, Chi and Quantum Consciousness.* Tucson: Poseidia Press.
23. Cho, Z. H., S. C. Chung, J. P. Jones, et al. 1998. "New Finding of the Correlation between Acupoints and Corresponding Brain Cortices using Functional MRI." *Proceedings of the National Academy of Science* 95: 2670–73.
24. Jones, J. P., and Young K. Bae. 2004. "Ultrasonic Visualization and Stimulation of Classical Oriental Acupuncture Points." *A Journal for Physicians by Physicians* 15 (2): 24–26.
25. Jones, Joie P., and Young K. Bae. 2008. "The Imaging of Acupuncture Points and the Characterization of Signal Pathways using fMRI and Quantitative Ultrasonic Methods." *Journal of Society for Scientific Exploration.*
26. Bruyere, Rosalyn L. 1989. *Wheels of Light: A Study of the Chakras.* Sierra Madre: Bon Productions.

Chapter 3. Atoms and Strings: Access to Subtle Energy

1. Phillips, Stephen M. 1999. *ESP of Quarks and Superstrings.* New Delhi, India: New Age International (P) Limited.
2. Ginzburg, Vladimir B. 2006. *Prime Elements: Of Ordinary Matter, Dark Matter & Dark Energy.* Pittsburgh, Pa.: Helicola Press.
3. Greenberg, O. W. 1964. "Spin and Unitary—Spin Independence in a Paraquark Model of Baryons and Mesons." *Phys. Rev. Lett.* 13: 598–602.
4. Leadbeater, C. W., and Annie Besant. *Occult Chemistry, Investigations by Clairvoyant Magnification into the Structure of the Atoms of the Periodic Table and Some Compounds.* Third Edition. Kila, Mont.: Kessinger Publishing Company.
5. Zuyin, Lu. 1997. *Scientific Qigong Exploration: The Wonders and Mysteries of Qi.* Malvern, Pa.: Amberleaf Press.
6. Swanson, Claude. 2010. *Life Force, the Scientific Basis: Breakthrough Physics of Energy Medicine, Healing, Chi and Quantum Consciousness.* Tucson: Poseidia Press.
7. Smrz, Milan. 1988. "Experiments Made Together with Robert Pavlita." The Seventh International Conference on Psychotronic Research. Carrollton, Georgia: West Georgia College, B-18.

Chapter 4. Vital Force Technology

1. Halevi, Z'ev ben Shimon. 1977. *A Kabbalistic Universe.* Boston, Mass./ York Beach, Maine: Weiser Books.
2. Leadbeater, C. W., and Annie Besant. 1950. *Occult Chemistry, Investigations by Clairvoyant Magnification into the Structure of the Atoms of the Periodic Table and Some Compounds.* Kila, Mont.: Kessinger Publishing Company.
3. Phillips, Stephen M. 1980. *Extra-Sensory Perception of Quarks.* Chicago: University of Chicago Printing Department.
4. Jahn, Robert G., and Brenda J. Dunne. 1987. *Margins of Reality: The Role of Consciousness in the Physical World.* San Diego, New York, London: Harcourt Brace Jovanovich.
5. Zuyin, Lu. 1997. *Scientific Qigong Exploration—The Wonders and Mysteries of Qi.* Malvern, Pa.: Amber Leaf Press.
6. Radin, Dean. 1997. *The Conscious Universe: The Scientific Truth of Psychic Phenomena.* New York: HarperCollins.
7. Tiller, William A., Walter E. Dibbie, Jr., J. Gregory Fandel. 2004. *Some Science Adventures with Real Magic.* Walnut Creek, Calif.: Pavior Publishing.
8. Lipton, Bruce. 2005. *The Biology of Belief: Unleashing the Power of Consciousness, Matter and Miracles.* Santa Rosa, Calif.: Mountain of Love.
9. Radin, Dean. 2007. *Entangled Minds: Extrasensory Experiences in a Quantum Reality.* New York: Paraview Pocket Books.
10. Tiller, William A. 2007. *Psychoenergetic Science: A Second Copernican-Scale Revolution.* Walnut Creek, Calif.: Pavior Publishing.
11. Radin, Dean. 2013. *Supernormal: Science, Yoga, and the Evidence for Extraordinary Psychic Abilities.* New York: Deepak Chopra Books.
12. Chen, Kevin W. 2004. "An Analytic Review of Studies on Measuring Effects of External Qi in China." *Alternative Therapies,* July/August: 38–50.
13. Church, Dawson. 2018. *Mind to Matter: The Astonishing Science of How Your Brain Creates Meterial Reality.* Carlsbad, Calif.: Hay House.
14. Lesniak K. T. 2006. "The effect of intercessory prayer on wounds healing in nonhuman primates." *Alternative Therapies in Health and Medicine,* Nov.–Dec., 12 (6): 42–48.
15. Rao, Manju Lata, Tania M Slawecki, M. Richard Hoover, and Rustam

Roy. 2008. "Characterization and Properties of Structured Waters." "Materials Day" at the Materials Research Institute, Penn State University. University Park, Pa.
16. Swanson, Claude. 2010. *Life Force, the Scientific Basis: Breakthrough Physics of Energy Medicine, Healing, Chi and Quantum Consciousness.* Tucson: Poseidia Press.
17. Tania M Slawecki. 2007. "Future Energy and Future Medicine." *USPA Proceedings:* 222–28.
18. Schwartz, S. A., De Mattei, R. J., Brame, E. G., & Spottiswoode, J. P. 1990. "Infared Spectra Alteration in Water Proximate to the Palms of Therapeutic Practitioners." *Subtle Energies,* vol. 1 (1): 43–72.

Chapter 5. Vital Seeds, Healthy Plants

1. Levengood, W. C., and Yury Kronn. 2002. "Monoatomic Subtle Energy Patterns Induced in Water and their Influence on Plant Growth." 21st Annual Conference for the Society for Scientific Exploration. Charlottesville, Virg.

Chapter 6. Energy Medicine

1. Klimavičiusa, Linda, K. Jekabson, John McMichael, and Yury Kronn. 2014. "Effect of Subtle Energy Patterns on Cell Viability and Mitochondrial Membrane Potential." 33rd Annual Conference of the Society for Scientific Exploration. San Francisco.
2. Lipton, Bruce. 2005. *The Biology of Belief: Unleashing the Power of Consciousness, Matter and Miracles.* Santa Rosa: Mountain of Love/ Elite Books.
3. Lins, Jeremy. 2014. "Effect of Subtle Energy Patterns on Gene Regulation of Human." Society for Scientific Exploration, 33rd Annual Conference. San Francisco.

Chapter 7. The Unseen Enemy: Energetic Pollution

1. Swanson, Claude V. 2009. *Life Force: The Scientific Basis. Breakthrough Physics of Energy Medicine, Healing, Chi and Quantum Consciousness.* Tucson: Poseidia Press.

2. Zuyin, Lu. 1997. *Scientific Qigong Exploration—The Wonders and Mysteries of Qi*. Malvern, Pa.: Amber Leaf Press.
3. Jones, Joie., Yury Kronn. 2008. "New Understandings on the Effects of Energetic Pollution on the Healing Process and Solutions Made Possible with Modern Subtle Energy Technology." *Medical Week*. Baden-Baden, Germany.
4. Kronn Y. 2019. "Energetic Pollution—The Unseen Enemy. Can Frontier Science Counteract It?" *Acta Scientific Pharmaceutical Sciences,* vol. 3, no. 3.
5. Nazarov, Igor. 2019. Schumann Resonance, Brainwaves, Neuro-feedback and Beyond. Self-published. Accessible via ResearchGate or Vital Force Technology websites.

Appendix A. Electrons, Subquarks, and Superstrings Examined by Stephen M. Phillips, Ph.D.

1. Phillips, S. M. 1999. *ESP of Quarks and Superstrings*. New Delhi, India: New Age International (P) Limited.

Appendix B. Fibonacci Sequences of the Traditional Chinese Medicine Elements

1. Chen, Zhaoxue. 2015. "Researches on Mathematical Relationship of Five Elements of Containing Notes and Fibonacci Sequence Modulo 5." *The Scientific World Journal* ID 189357.
2. Unschuld, P. 2003. *Huang Di Nei Jing Su Wen: Nature, Knowledge, Imagery in an Ancient Chinese Medical Text*. Berkeley: University of California Press.

Appendix D. Phantom Atom Effects on Animals and Human Cells

1. Marrongelle, Jeffrey, and Yury Kronn. 2002. "The Effect of Subtle Energy on Autonomic System Response as Quantified by the Heart Rate Variability Test." *Newsletter of the Society for Scientific Exploration*. Vol. 18, nos. 1, 2, 3.

2. Svirskis, Simons, Linda Klimavičiusa, and Zane Dzirkale. 2018. "Evaluation of 'Stress Relief' dietary supplement on animal stress level and locomotion." *Proceedings of the Latvian Academy of Sciences* DOI: 10.2478/prolas-2018-0027.
3. Klimavičiusa, Linda, K. Jekabson, John McMichael, and Yury Kronn. 2014. "Effect of Subtle Energy Patterns on Cell Viability and Mitochondrial Membrane Potential." 33rd Annual Conference of the Society for Scientific Exploration. San Francisco.

Appendix E. Pranic Healing and Electromagnetic Pollution

1. Marrongelle, Jeffrey, and Yury Kronn. 2002. "The Effect of Subtle Energy on Autonomic System Response as Quantified by the Heart Rate Variability Test." *Newsletter of the Society for Scientific Exploration*. San Francisco. Vol. 18, nos. 1, 2, 3.

Acknowledgments

THIS BOOK WAS MADE POSSIBLE due to the twenty long years of work by my wife, Constance, who as the president of our Energy Tools Int., LLC, put tremendous efforts into finding possibilities for scientific research of the energy formulas we had been creating. She left this world too early, and my book is dedicated to her blessed memory.

During all the years of developing and using my Vital Force Technology, many people with extraordinary medical intuitive abilities provided me with invaluable assistance in perfecting the technology to create crystal-clear energy patterns without any viruses. Among them are Pam Sheffer, Galina Kalyuzhny, Laura E. Graye, and several others who did not want their names to be known.

I express my deep appreciation to William Mathes and Jurriaan Kamp, who made this book easily readable.

I am very thankful to my colleagues Dr. Simons Svirskis, Dr. Igor Nazarov, and Andrei Kogan, who helped me to present the experimental results in a format appropriate for this book.

Index